ENERGY AND CLIMATE

Energy and Climate

VISION FOR THE FUTURE

Michael B. McElroy

Oxford University Press is a department of the University of Oxford. It furthers
the University's objective of excellence in research, scholarship, and education
by publishing worldwide. Oxford is a registered trade mark of Oxford University
Press in the UK and certain other countries.

Published in the United States of America by Oxford University Press
198 Madison Avenue, New York, NY 10016, United States of America.

© Oxford University Press 2016

All rights reserved. No part of this publication may be reproduced, stored in
a retrieval system, or transmitted, in any form or by any means, without the
prior permission in writing of Oxford University Press, or as expressly permitted
by law, by license, or under terms agreed with the appropriate reproduction
rights organization. Inquiries concerning reproduction outside the scope of the
above should be sent to the Rights Department, Oxford University Press, at the
address above.

You must not circulate this work in any other form
and you must impose this same condition on any acquirer.

Cataloging-in-Publication data is on file at the Library of Congress
ISBN 978-0-19-049033-1

9 8 7 6 5 4 3 2 1
Printed by Sheridan, USA

For my brothers Eamon and Joseph and the memory of our parents

Contents

Preface ix
Acknowledgments xiii

1. Introduction 1

2. Energy Basics 13

3. The Contemporary US Energy System: Overview Including a Comparison with China 25

4. Human-Induced Climate Change: Why You Should Take It Seriously 34

5. Human-Induced Climate Change: Arguments Offered by Those Who Dissent 65

6. Coal: Abundant but Problematic 79

7. Oil: A Volatile Past, an Uncertain Future 89

8. Natural Gas: The Least Polluting of the Fossil Fuels 107

9. Nuclear Power: An Optimistic Beginning, a Clouded Future 123

10. Power from Wind: Opportunities and Challenges 143

11. Power from the Sun: Abundant but Expensive 160

12. Hydro: Power from Running Water 178

13. Earth Heat and Lunar Gravity: Geothermal and Tidal Energy 193

14. Plant Biomass as a Substitute for Oil in Transportation 205

15. Limiting US and Chinese Emissions: The Beijing Agreement 219

16. Vision for a Low-Carbon-Energy Future 234

Index 263

Preface

THE CLIMATE OF our planet is changing at a rate unprecedented in recent human history. The energy absorbed from the sun exceeds what is returned to space. The planet as a whole is gaining energy. The heat content of the ocean is increasing, the surface and atmosphere are warming, mid-latitude glaciers are melting, sea levels are rising, and the Arctic Ocean is losing its ice cover. All of these assertions are based on hard observational facts.

Increases in the concentrations of greenhouse gases, notably carbon dioxide, methane, and nitrous oxide, are largely responsible for the imbalance in the energy budget of the planet. Combustion of coal, oil, and natural gas is primarily responsible for the rise in the concentrations of these gases, which are higher now than at any time over at least the past 800,000 years. If we are to arrest the current trend, we need to switch to an alternate energy economy. The challenge to respond is immediate. The impacts of a changing climate are already apparent and can only increase in disruptive potential in the future.

It is important to distinguish between climate and weather. The climate experienced in any particular location reflects the average long-term properties of the weather. Weather by its very nature is variable, essentially unpredictable on time scales longer than about 2 weeks. There is persuasive evidence that weather has become increasingly weird, as *New York Times* columnist Thomas Friedman described it, in recent years—more floods, more droughts, more violent storms, and more extreme heat waves. This is precisely what one would expect from a warming climate. The water vapor content of the atmosphere is increasing. Condensation of water vapor, which releases heat, is the primary energy source for the atmosphere. If meteorological conditions favor rising motion heralding an approaching storm, given the higher concentration of water vapor in the atmosphere, it is likely to rain or snow more than would be the case otherwise. It follows that if the supply of moisture to the atmosphere is to remain relatively constant on a global scale (a plausible assumption), it must rain less elsewhere—more floods, more droughts. More energy to fuel storms increases the probability that the storms when they develop will pack a greater punch. Persistent drought in a forested region increases the likelihood of fires that can be triggered either innocently or deliberately by human agents or naturally by lightning. In either case, the damage to property, human lives, and natural resources can be serious and extensive.

There is a consensus in the informed scientific community that the threat of imminent, human-induced climate change is real and should be taken seriously. Yet there are those, particularly in the United States, who dispute this conclusion. With few exceptions, the naysayers, while distinguished in other fields, are not climate scientists. Their views, as I argue later in this volume, reflect personal opinions, uninformed by the observational data and basic science that are used by professional climate scientists, but nonetheless influential. They demand a response. Accordingly, I elected to include two chapters in this book, the first outlining reasons why one should take the climate issue seriously, the second responding to the criticisms advanced by those who dissent.

For the past several years I have taught a course at Harvard University under the auspices of the university's General Education curriculum. The intent of this program is to expose students to modes of thinking and analysis exercising skills judged essential for a broadly based liberal arts education. The General Education program is designed to complement expertise students acquire in their more specialized fields of study—in traditional disciplines such as the physical and biological sciences, the social sciences, and the humanities. The students who elect to take the course I offer, which focuses on energy and the environment, come from a variety of disciplinary backgrounds. My objective is to provide a broad-based introduction to the topics covered in the course, ranging from the historical to the immediate to the future. A primary purpose for my earlier book (*Energy: Perspectives, Problems, and Prospects*: Oxford University Press, 2010) was to support the material presented in this course. While intended to be accessible to a broader audience, the book included equations and numerical examples. A number of my nonscientist friends lamented the fact that they had to suffer through this level of detail, suggesting that it would be more effective and reach a larger audience if the book could be reissued with these technical details deleted.

I had this advice in mind when I set out to write the present book. So I made the decision to use words and figures to present the relevant material rather than equations and numerical examples. It was important, however, not to compromise on rigor. To this end, I chose to include references to primary literature for those who might wish to delve more deeply into the topics discussed. Understanding climate and related energy issues is critical, I believe, if we are to have an informed debate on these issues and if we are to responsibly address the underlying challenge. I am dismayed by the lack of content in much of the public discussion of these issues in the United States. My hope is that the material presented here should provide at least background to encourage informed debate. The general reader may find the extensive compilations of data on energy systems and the rather detailed discussion of the climate system a little off-putting, but I believe that it is important to present the facts and not gloss over the details. Exposure to a typical high school science curriculum should provide more than sufficient background to allow the interested reader to follow the presentation. If the reader does not wish to peruse the details presented in the multiple figures and tables of the book, it should be relatively easy to skip on and follow the main story.

The book begins with a general introduction, followed by chapters on energy basics, a discussion of the contemporary energy systems of the United States and China, and the aforementioned chapters on climate. It continues with a series of chapters on specific energy options: coal, oil, natural gas, nuclear, wind, solar, hydro, geothermal, and biomass. The perspective is global but with a specific focus on the United States and China, recognizing the critical role these countries must play in addressing the challenge of global climate change. China now ranks number one in terms of emissions of the key greenhouse gases, notably carbon dioxide, having recently supplanted the United States. The book concludes with a discussion of initiatives now underway to at least reduce the rate of increase of greenhouse gas emissions, together with a vision for a low-carbon-energy future that could in principle minimize the long-term impact of energy systems on global climate. We, society, need to transition to a future based to a large extent on renewable—low-carbon—sources of energy such as wind, sun, hydro, biomass, geothermal, and potentially nuclear. We know how to do it. To succeed, as I discuss, will require focused investments in infrastructure, notably enhancement of existing networks for distribution of electric power, implementation of anticipated advances in technology, global cooperation, and above all, committed political leadership.

Acknowledgments

IN LATE AUGUST of 2015, large regions of the Western United States were suffering from out-of-control wildfires, an inevitable result of the extended drought the region has experienced over the past several years. Lives have been lost, property has been devastated, and the economic damage is estimated in the billions of dollars. Firefighters have been imported from as far away as Australia and New Zealand. Weather has become increasingly extreme, in terms not only of precipitation and temperature but also in the violence of storms. And the impacts have not simply been confined to the Western United States: they extend to the globe as a whole.

The underlying cause is not a mystery: it relates directly to the buildup of heat-trapping gases in the atmosphere, a consequence primarily of our excessive, unsustainable reliance on carbon dioxide–emitting sources of coal, oil, and natural gas as the primary inputs to our modern energy economy. Levels of carbon dioxide in the atmosphere are higher now than at any time over at least the past 850,000 years. They are destined, unless we take action, to climb to levels not seen since dinosaurs roamed the planet more than 50 million years ago. Our planet is absorbing more energy from the sun than it is returning to space, with the excess stored primarily in the ocean. Patterns of weather are changing. Sea levels are rising. As discussed in this book, there is little question that we are responsible for these changes.

Reaction to the challenge posed by these issues, particularly in the United States, has been based too often not on consideration of the relevant science but rather on positions fixed by allegiance to particular political philosophies. A central factor in my decision to write this book was the conviction that responses to the issue of human-induced climate change should be set by objective analysis of the relevant facts. Accepting that there is a problem should be the starting point for a constructive dialogue on how the issue should be addressed. My objective has been to define the challenge and to provide a conceptional vision of how to tackle it.

The book includes two chapters on climate, the first presenting arguments as to why one should take the issue of human-induced change seriously, the second addressing objections posed by some who take a different view. The material included in the first of these chapters was informed by a study that D. James Baker and I conducted a few years ago for the US Intelligence Community on the implications of climate extremes for national security (http://environment.harvard.edu/sites/default/files/climate_extremes_report_2012-12-04.pdf). My thanks to Linda Zall for her foresight in commissioning this study and to the 10 distinguished experts (identified by name in the report) who served as reviewers. A particular thanks to Gregory Bierman, Ric Cicone, Marc Levy, and Tom Parris for sharing their expertise and for the valuable contributions they made to the study.

I am indebted to a number of colleagues—D. James Baker, Ralph Cicerone, Xi Lu, Chris Nielsen, Ruth Reck, William Schlesinger, and Yuk Yung—who took the time to read the entire manuscript. My heartfelt thanks to them for their unselfish, time-consuming contributions to enhancing the quality of the ultimate product. A special thanks to Xi Lu. In working on this project, I benefitted immeasurably from Xi's deep and insightful understanding of all aspects of energy science. My thanks also to Junling Huang for sharing his enthusiasm for science and for the excellent PhD thesis he wrote under my direction.

I have had the privilege over my career at Harvard to associate with and learn from a large number of distinguished colleagues. A special thanks to Fred Abernathy, George Baker, Jim Anderson, Daniel Jacob, Dale Jorgenson, Dan Schrag, and Steve Wofsy, who shared my interests on the future of energy. Conversations and email exchanges with Harvard Business School professor Joseph Lassiter prompted me to take a more nuanced position on the future of nuclear energy. Thanks, Joe.

I have had the honor and pleasure over the past several years to advise and work with a number of talented undergraduates at Harvard, who elected to write senior theses under my direction on a range of energy-related subjects. I would single out in particular Andrew Cohen, Charles Gertler, Jonathan Park, Jackson Salovaara, Jun Shepard, Nora Sluzas, and Jeremy Tchou (names listed in alphabetical order). Their work was invariably of high quality, recognized in several instances by distinguished awards at the university and in some cases by subsequent publications in the professional literature. My thanks to them for the enthusiasm they brought to their work and for the many important contributions they made to our research program.

Finally, I would like to express my debt to Stephen McElroy, who read large portions of the manuscript and made suggestions that markedly improved its content and presentation, and to Cecilia McCormack, who continues to maintain a level of order in my office, my research group, and my professional life.

1

Introduction

THE RISK OF disruptive climate change is real and immediate. A low-pressure system forming in the tropics develops into a Category 1 hurricane, making its way slowly up the east coast of the United States. Normally a storm such as this would be expected to make a right-hand turn and move off across the Atlantic. Conditions, however, are not normal. This storm is about to encounter an intense low-pressure weather system associated with an unusual configuration of the jet stream, linked potentially to an abnormally warm condition in the Arctic. Forecasts suggest that rather than turning right, the storm is going to turn left and intensify as it moves over unseasonably warm water off the New Jersey coast. It develops into what some would describe as the storm of the century. New York and New Jersey feel the brunt of the damage. The impact extends as far north as Maine and as far south as North Carolina. Lower Manhattan is engulfed by a 14-foot storm surge, flooding the subway, plunging the city south of 39th Street into darkness. Residents of Staten Island fear for their lives as their homes are flooded, as they lose power, and as their community is effectively isolated from the rest of the world. As many as 23 people are drowned as floodwaters engulf much of the borough. Beach communities of New Jersey are devastated. As much as a week after the storm has passed, more than a million homes and businesses in New York and New Jersey are still without power. Estimates of damage range as high as $60 billion. This is the story of the devastation brought about by Hurricane Sandy in late October of 2012.

The encounter with Sandy prompted a number of queries concerning a possible link to human-induced global climate change. Andrew Cuomo, governor of New York, commented: "Part of the learning from this is the recognition that climate change is a reality, extreme weather is a reality." Mayor Michael Bloomberg's reaction was more nuanced: "Our climate is changing. And while the increase in extreme weather we have experienced in New York City and around the world may or may not be the result of it, the risk that it might be—given this week's devastation—should compel all elected leaders to take immediate action." President Obama, speaking following his election to a second term as president, stated: "We want our kids to grow up in an America . . . that isn't threatened by the destructive power of a warming planet." There were others, though, who doubted the connection. A reporter on the business channel CNBC offered the opinion that the hurricane should simply be considered bad luck—a one-in-a-hundred-year event. A guest on the same program suggested that we could now relax: we were unlikely to witness a repeat for a hundred years in the future. How can we reconcile these disparate views? What are the facts?

The threat of serious climate change is real. Sea level is rising—faster than predicted. If as little as 10% of the Greenland ice sheet were to collapse, that would be good for as much as a meter rise in *global* sea level. A large portion of the Greenland ice sheet was covered by a layer of liquid water for several days during the summer of 2012, a result of the exposure of the ice sheet to unusually warm weather. Collapse of the ice sheet in West Antarctica, currently grounded on land, would be even more serious. The Arctic Ocean is losing its summer ice cover: prospects are that this region will be largely ice-free for at least part of the year within a few decades. Weather is increasingly more extreme—we are experiencing more record-breaking heat waves, devastating forest fires, violent storms, floods, and droughts. And it is likely to get much worse if we fail to act promptly to deal with the underlying cause: the inexorable rise in the concentration of so-called greenhouse gases, most important, the large quantities of carbon dioxide (CO_2) we are adding to the atmosphere when we burn coal, natural gas, and oil, the primary sources of energy for our modern industrial economy. The concentration of CO_2 is higher now than at any point over at least the past 800,000 years. Given current trends, it is likely to rise over the next few decades to levels not seen since dinosaurs roamed the Earth 50 million years ago. And CO_2 is not the only climate-impacting greenhouse gas we have to worry about. Concentrations of methane (CH_4) and nitrous oxide (N_2O), even more potent greenhouse agents than CO_2, are also at record high levels and likely to increase further in the future.

Can we be certain that there is a direct connection between storms like Hurricane Sandy and global climate change? The answer is no. Sandy may have been simply a bad roll of the dice—a one-in-a-hundred-year happenstance as could be inferred from a literal reading of the historical record. But the odds for a repeat are increasing: what was once a one-in-a-hundred-year event is likely to become more common in the future (one in 20 years, one in 10 years?). The ground rules for the weather/climate game are changing. What happened in the past is not a reliable guide to what may develop in the future.

Continuing to rely indefinitely on climate-impacting fossil fuels—coal, oil, and natural gas—is not an acceptable option. We need to transition as quickly as possible to a more acceptable, more sustainable trajectory to meet our future demand for energy. And we need to do so while at the same time minimizing the potential damage to our economy, to the way of life we have come to take for granted. This is a challenge that can be met. But if we are to address it successfully, we must first accept that the problem is real.

The first step is to understand the issue fully and then to confront those who would seek to cast it as ideological or use it to achieve short-term political or economic advantage. My objective here is to offer an honest presentation of the facts—what we know and what we do not know—to contribute hopefully to a more informed discussion of what is admittedly a complicated topic. I propose to outline the reasons why it should be taken seriously. I will also present and comment on the arguments advanced by those who would take a different view. The question of whether the threat of human-induced climate change is real or not, and how serious it may be, should not be treated as an issue for political gamesmanship. However, politics can, and indeed should, play a role in a discussion of how we might propose to deal with it.

Clearly, there is scope for debate on how best to achieve this objective. But we need to begin an intelligent discussion. Putting it off until tomorrow, or to the next electoral cycle, is not a responsible option.

LIFE IS PERSISTENT: HUMANS ARE VERY RECENT

It is instructive to understand the pervasive influence that life has had on the history of our planet and just how recently humans have come to assume the dominant role we occupy today.

The Earth is approximately 4.5 billion years old. Life has been a feature of our planet's history for at least the past 3.5 billion years. The earliest forms of life consisted of organisms known as prokaryotes, simple creatures with the ability to draw energy either from chemicals in their environment or directly from sunlight. Bacteria and blue green algae are examples of these early life forms, and they have been remarkably resilient. They continue to play an important role in the complex web that characterizes life on our planet today. Stage 2 in the evolution of life involved the appearance of what are known as the eukaryotes, unicellular organisms significantly more complex than their prokaryotic antecedents. The late Lynn Margulis suggested that the eukaryotes evolved through the fusion of cells of preexisting prokaryotes—think of one species of prokaryotes consuming (eating) another, thus incorporating and combining their DNA. The pace of biological evolution has been distinctly sporadic. There were periods when large numbers of new species appeared, offset by times characterized by massive extinctions. Progenitors of essentially all of the modern phyla are present, for example, in the Burgess Shale, a remarkable assemblage of fossils, dated at approximately 440 million years before present (BP), discovered by C. D. Walcott high in the Canadian Rockies in 1909.

The explosion of new life forms that appeared during the Cambrian Period (543 to 510 million years BP) as recorded in the Burgess Shale was followed a few hundred million years later (225 million years BP) by what the late Stephen Jay Gould described as the "granddaddy of all extinctions." This momentous event resulted in the elimination of as much as 95% of all of the marine species that were alive at that time. A second major extinction 65 million years BP, triggered in this case by the impact of a giant meteorite (or meteorites), led to the demise of the dinosaurs, opening up a niche for the later ascendancy of large mammals and eventually humans. As Gould drolly remarked: "In an entirely literal sense, we owe our existence, as large and reasoning animals, to our lucky stars." There can be little doubt that large-scale environmental change played a major role in these epochal extinction events. On the flip side, elimination of maladjusted species contributed undoubtedly to opening up opportunities for new species to step in and take their place.

Plate tectonics have contributed in no small measure to the long-term persistence of life on Earth. As organisms die, in the ocean for example, a fraction of their body parts fall to the sea bottom, where they can be buried and effectively immobilized in sediments. This process is responsible for important removal of life-essential elements such as carbon, nitrogen, and phosphorus. Consider carbon, for example. The lifetime of carbon in the combined atmosphere-ocean-biosphere, the time it takes on average before the typical carbon atom is transferred to the sediment, is less than 200,000 years. What this means is that if the loss of carbon to the sediments were permanent, the surface environments of the Earth (the regions important for life) would long ago have run out of carbon, an essential component of all living organisms. Fortunately for us, the sedimentary sink is transitory. Sediments are transported on the giant crustal plates that float like rafts on the hot heavier material of the underlying mantle. When plates converge, they can be either uplifted or they may be withdrawn into the underlying mantle. Uplift results in the formation of mountain ranges. Carbon (and other life-essential elements) may be cycled directly back to the atmosphere/biosphere system, in this case by weathering of the uplifted rock material. If the sedimentary material is carried down into the mantle, it will be raised to high temperature through exposure to the hot mantle material. The carbon and other volatile materials included in the sediments may be released and transferred, often explosively, back to either the atmosphere or ocean as a component of hot springs and volcanoes. The average carbon atom has gone through this tectonically driven cycling sequence at least 10 times over the course of Earth history.

How do humans fit into this grand saga of biological evolution? The first sobering realization is just how recently it is that we made our entry onto the stage of life: studies of mitochondrial DNA suggest that we have a common maternal ancestor, a primitive Eve, who lived in Africa some 150,000 years BP. Our earliest ancestors were nomads—hunter-gatherers—constantly on the move in search of food. By 20,000 years BP, they had extended their reach to encompass all of the world's continental landmasses with the exception of Antarctica. The first humans to arrive in the Americas probably did so about 30,000 years BP, making their way across what is now the Bering Straits, taking advantage of the fact that sea level then, during the last great ice age, was approximately 120 meters lower than it is today (North America and Asia were joined at that time by a land bridge).

Think of what this means. If we were to compress the 4.5 billion-year history of the Earth into a single year, it was not until 3.5 minutes before midnight on December 31 that the first humans made their way to the Americas!

The first transformative event in the history of our species involved arguably the development of agriculture and animal husbandry. It is likely that this took root first in the Middle East, about 10,000 years BP, in the region known as the Fertile Crescent. Our ancestors learned subsequently to fashion tools and weapons from combinations of copper and tin and later iron (all this between about 5,000 and 3,000 years BP). Access to affordable energy, mainly wood in these early days, was critical to the success of the various civilizations that developed subsequently. When the wood supply ran out, civilizations collapsed. It was a pattern repeated many times throughout history—in Mesopotamia, Egypt, the Indus region, Babylon, Crete, Greece, Cyprus, and later in Rome—as discussed, for example, by Perlin (1989).

Early Western civilizations clustered around the Mediterranean, shifting later to locations along the Atlantic seaboard, notably Spain and Portugal initially, later France, Holland, and England. Throughout this several thousand year-long interval, civilization in the east would be dominated by China. Indeed, as Fairbank and Goldman (2006) pointed out, for much of this time China was "the superior civilization of the world, not only the equal of Rome but far ahead of medieval Europe." From the 1500s to the early 1800s, China's economy was the largest in the world. The modern industrial age would dawn, however, not in China but on the northwestern periphery of Europe, in England. It developed approximately 250 years ago (2 seconds before midnight on December 31 in terms of our earlier analogy). The world would never be the same.

THE INDUSTRIAL REVOLUTION

The events that unfolded in England over the course of the late eighteenth and early nineteenth centuries are referred to commonly as the Industrial Revolution. They involved for the most part the transition from a world where work was performed primarily by human and animal muscle power to one dominated by machines energized either by running water or, especially in the later years, by steam generated by burning coal. It is important to understand just how epochal these developments were. They provide essential context for the challenges we face today.

Consumer goods produced efficiently in England found a ready market not only in England but across a large swathe of the world taking advantage of the country's mercantile tradition and the extensive reach afforded by her widespread colonial possessions. The opportunities for creation of wealth afforded by the new age of mechanization were quickly adopted by England's neighbors on continental Europe and in the New World, in the United States. They spawned a series of innovations that literally transformed the world.

The steam engine introduced in England by Thomas Newcomen in 1712 as a means to pump water from underground mines was perfected by James Watt in the latter half of the eighteenth century and applied subsequently to grind flour, to run cotton mills, to fuel the locomotives of the early railroads, to replace sail as the motive force for shipping, and for a host of other applications. Samuel Morse, capitalizing on the emerging understanding of electricity, introduced the first commercially available telegraph in 1843, successfully implementing the first long-distance exchange of telegraph messages between the terminal of the Baltimore and Ohio (B and O) railroad in Baltimore and the Supreme Court Chamber in Washington, DC. Within 5 years there were more than 12,000 miles of telegraph lines in the United Sates run by no fewer than 29 different companies. The first transatlantic cable went into operation in 1866. Alexander Graham Bell invented the telephone, and the first telephone exchange was established in New Haven in 1878. Thomas Alva Edison, the most prolific inventor of his time, set up what was arguably the world's first organized research laboratory at Menlo Park in New Jersey, perfected the light bulb, and established the first central power station for production and distribution of electricity (on Pearl Street, in Lower Manhattan, New York) in 1882. Henry Ford introduced the concept of an assembly line with distributed labor responsibilities and associated efficiencies and capitalized on this innovation to produce in 1908 the first widely affordable automobile, the Model T. By 1927, when the last car rolled off the assembly line, 15 million Model T's had been sold worldwide in no fewer than 21 different countries. Much of this impressive growth was fueled by access to abundant sources of affordable energy—starting with coal, progressing to oil, and more recently to natural gas. The unintended consequence was a precipitous rise in the concentration of atmospheric CO_2.

THE ACCELERATING PACE OF CHANGE

It is important to understand just how different the world is today as compared to the one that existed in the immediate aftermath of World War II. As Thomas Friedman perceptively observed, we are living now in a world that is increasingly flat. If it is cheaper to produce consumer goods in China or India and to sell them in Europe or the United States, and if there is profit to be made by doing so, smart business will assuredly act to take advantage of this opportunity. Thanks to instant communications, a shortfall in grain production induced by unfavorable weather in Russia or Argentina can have an essentially immediate impact on the price of grain—and bread—in regions far removed from the original shortfall. Stock markets rise and fall globally in sync. Carbon dioxide emitted from a factory producing steel in China for export to the United States will have an impact not just on climate in both of these countries but also on climate worldwide. Such are the facts of life in our increasingly interconnected, interdependent world. We need to appreciate the significance of this unprecedented change and plan accordingly to prepare for conditions that will be assuredly very different in the future from those experienced by our parents and grandparents.

The population of the world is greater now than it has been ever in the past. There were approximately 2.5 billion people in the world in 1950. Today there are more than 7 billion.

Given current projections, we can expect more than 9 billion by 2050. Accounting for inflation, the world economy (the gross world product or GWP) has grown eight-fold since 1950. What made this possible was the development of a global market for goods and services, aided and abetted by advances in technology (in particular communications), ready access to energy (particularly the aforementioned fossil sources), and not least on the ease with which goods could be transported inexpensively and profitably over large distances. While growth and prosperity in the pre–World War II period were largely confined to countries in the West, the market has now expanded, reaching just about all parts of the inhabited world with the possible exception of distressed portions of Africa.

Japan and the so-called Asian Tigers—South Korea, Taiwan, Hong Kong, and Singapore—were early beneficiaries of the expansion in global trade, joined more recently by China and India, the world's two most populous countries. China's modern economic growth began in the late 1970s, triggered by the reforms initiated by Deng Xiaoping. India took off a decade or so later. Now the economies of both countries are growing at annual (percentage) rates approaching double digits: 7.7% for China in 2013 (a decrease from 10.4% in 2010) and 5.0% for India in the same year (a decrease from 10.3% in 2010). This compares with an overall growth of 1.9% for the United States in 2013, −0.4% for the 15 countries of the European Community, and 3.1% for the world as a whole (all of these data for 2013). The world is richer now than it has ever been in the past. At the same time, disparities in incomes have increased both within and across countries.

A CRITICAL CHALLENGE

HSBC, the international banking company, projected in 2012 that the global economy would increase by as much as a factor of 3 by 2050. Implicit in its analysis was the assumption that there will be no surprises. The current trend is not sustainable. We need to be able to anticipate problems, not simply to react when they occur. Can we realistically expect to triple the size of the global economy in less than 40 years without straining the life-essential ecological resources of the planet? Can we develop the energy resources that will be required to enable this growth without seriously disrupting the climate system that supported our past success? Can we ensure continuing stability of the political entities required to maintain a level of fairness and order in our increasingly interconnected world? It is likely that we will have to deal with all of these complex, interrelated issues over the next few years. We need to anticipate the problems and be prepared.

The demand for food to feed our burgeoning population poses an important challenge for the future success of the human enterprise. We have increased markedly over the past half-century the quantity of food we produce per unit of cultivated land area. We succeeded in doing this by applying increasing concentrations of pesticides, herbicides, and chemical fertilizers to our fields, drawing water from subsurface aquifers to irrigate them, and by investing in selective breeding to improve the yield of key crop species. All of this required energy, supplied mainly by carbon-rich fossil sources. As we have become richer, we have elected to move up the food chain, to turn increasingly to animal intermediaries for meat and dairy products rather than relying solely on the crops that we grow in our fields. This adds to the demand for food crops: our animals also have to eat.

A fraction of the nitrogen fertilizer we apply to our fields is converted to nitrous oxide (N_2O) and released to the atmosphere. Cows, sheep, and goats—animals in the ruminant family—produce methane (CH_4), as do the flooded paddies in which we produce rice. As indicated earlier, N_2O and CH_4 are both potent greenhouse gases and like CO_2 their concentrations in the atmosphere are also rising at an alarming rate. Technology can play an important role in addressing the important challenge of feeding what HSBC projects will be an increasingly affluent world. In doing so, though, we have to confront challenges posed by unanticipated shifts in weather and climate.

It is clear that the planet is under pressure also to keep up with the demands for environmental services, food, and energy for our growing and steadily more affluent global society. Pollution of air and water is impacting the health of large numbers of people, most notably in rapidly developing countries such as China and India. Deterioration of soils, depletion of groundwater, excessive heat waves, floods, and droughts are straining our ability to keep up with the demand for food. In an increasingly crowded world, we are unusually vulnerable to unanticipated natural disasters. Prices for essential energy commodities, especially oil, are high and volatile, exceptionally so in response to even minor disruptions in supply. Especially at risk are some of the poorest people on the planet, but even the relatively rich are not immune—witness the tolls taken by Hurricane Katrina in the United States in 2005 (1,800 deaths, $81 billion in reported damages), by the Tohuko Earthquake and subsequent tsunami in Japan in 2011 (61,000 deaths, economic impact estimated by the World Bank in this case at as much as $235 billion), and more recently by Hurricane Sandy. Underscoring all of these challenges is the prospect for changes in local, regional, and global climate, changes for which we are ill prepared.

WARMER AND COLDER CLIMATES IN THE PAST

A common argument advanced by some who would deny the evidence for human-induced climate change is that because climate changed in the past, what is occurring now is no different from what happened in the past. This view is based on an inadequate, indeed ill-informed, appreciation of the fact that there is no precedent for the volume and accelerating rate at which we are adding greenhouse gases to the atmosphere at present. Our actions are resulting in an entirely new climate regime. What follows here is a brief overview of the historical context with respect to past climate.

Evidence suggests that the early Earth was relatively warm, a surprising result since the output of energy from the sun would have been much less then than it is today. The paradox of a warm Earth in the presence of a low solar energy output is referred to in the climate literature as the faint sun paradox. The conventional resolution of this paradox invokes the presence of a variety of compounds in the early atmosphere capable of trapping heat that would otherwise have been released directly to space—a more efficient greenhouse. It is unclear exactly what these compounds were. They could have included higher concentrations of carbon dioxide (CO_2), or methane (CH_4), or conceivably a variety of other infrared-absorbing species. A plausible argument can be advanced to support the presence of a higher concentration of greenhouse agents in the primitive atmosphere. This could have arisen as a result of an enhanced source from comets striking the Earth during the late stages of its formation. A greater concentration of volatile rich (gassy) material in early volcanism could also have contributed. And a more limited supply of fresh continental material available for weathering, for removal of gases such as CO_2, could have played a further role.

This initial warm period was followed by cooling. The first evidence for continental glaciation appears in the geologic record at about 2.5 billion years BP. This was followed, on at least four occasions between 750 and 580 million years BP, by a remarkable series of climatic extremes in which the Earth appears to have been frozen over from pole to pole, a condition referred to by Joseph Kirschvink as Snowball Earth (1992). Paul Hoffman and colleagues (1998) proposed that the Snowball Earth phenomenon was triggered by a precipitous drop in the concentration of atmospheric CO_2, resulting most likely from a temporary decrease in tectonic (volcanic) activity. Levels of CO_2, they suggested, decreased to such an extent that even in the tropics, temperatures would have dropped below the freezing point of water. A thick layer of surface ice would have capped the ocean, effectively isolating the ocean from the atmosphere. The concentration of CO_2, however, would not have remained low indefinitely. Rather, it would have begun a slow ascent in response to a continuing input from continental (land-based) volcanoes. Eventually, levels of CO_2 would have built up to the point where the associated climatic warming would have caused the ocean's ice cover to dissipate. Warming would have been accelerated by evaporation from the relatively warm waters below the previously frozen surface. Renewed contact between the atmosphere and ocean would have resulted in important changes in the chemistry of both the ocean and the atmosphere. This is evidenced, for example, by an explosive growth in the production of carbonate minerals observed at this time in marine sediments (referred to by geologists as cap carbonates). Life would have had to respond in this case not only to rapidly varying conditions in the ocean but also to important changes on land.

Climates were relatively warm during the geological period referred to as the Cretaceous, which lasted from about 145 to about 66 million years BP. Warm conditions persisted to about 5 million years BP. The forest-tundra boundary extended at that time to latitudes as high as 82° N, some 2,500 km north of its present location, occupying regions of Greenland now permanently covered in ice. There were times when frost-tolerant vegetation was common in Spitsbergen (paleolatitude 79° N), when alligators and flying lemurs lived happily on Ellesmere Island (paleolatitude 78° N), and when palm trees, incapable of surviving even temporary frost conditions, could grow and survive in Central Asia. None of this, of course, would be possible today. The implication of these observations is that temperatures were relatively mild even in winter at high northern latitudes despite the fact that the input of energy from the sun would have been minimal or even totally absent during this season.

How can we account for these remarkable conditions? There have been a number of suggestions. There are good reasons to believe that concentrations of CO_2 were significantly higher then than they are at present but not necessarily much greater than they could be in the not-too-distant future if we continue to rely on fossil fuels—coal, oil, and natural gas—as the primary resources for our global energy economy. One suggestion is that the Hadley circulation that dominates climatic conditions in the tropics today—the circulation responsible for rain forests in the tropics where moisture-laden air rises and for deserts where it descends in the tropics having largely exhausted its supply of water vapor—might have extended to higher latitudes. A second possibility is that the concentration of water vapor in the stratosphere might have been much higher than it is today (the consequence of a very different climate in the troposphere), contributing to an unusually efficient greenhouse in the high-latitude winter regime. This would have allowed the high-latitude region to retain a significant fraction of the heat accumulated over the preceding summer (Imbrie and Imbrie 1986). A key question is whether these conditions could recur in the future.

The Earth has been cold more often than it has been warm over the past several million years. Great sheets of ice formed on the continents over this interval, mainly in North America and in Northwestern Europe. The ice sheets waxed and waned over time, largely in response to subtle variations in the orbital parameters of the Earth, impacting notably the tilt of the rotation axis (referred to as obliquity) and the position of the Earth as a function of season on its orbit around the sun (referred to as the precession of the equinoxes). The intensity of sunlight incident at any particular latitude at any particular time varies in tune with these changes—on a time scale of about 41,000 years responding to variations in obliquity, and on a time scale of about 22,000 years reflecting the changes associated with the precession of the equinoxes. Theory suggests that the ice sheets advance during times when the intensity of sunlight reaching them in summer is in a declining phase; they retreat when it is increasing (McElroy 2002). There have been eight major ice ages over the past 800,000 years lasting individually about 100,000 years. The last ice age drew to a close approximately 20,000 years BP. Sea level at that time was about 120 meters lower than it is today, reflecting the great mass of water that had been withdrawn from the ocean to supply the demand for water in the continental ice sheets. As recent as 7,000 years BP, sea level was still approximately 20 meters lower than it is today.

Recovery of climate from the last ice age was episodic. A sharp pulse of warming was observed both in Greenland and in the tropics beginning at about 15,000 years BP. This was followed by an abrupt climate reversal, a resumption of near glacial conditions that set in at about 13,000 years BP and lasted about 2,000 years. This cold snap, referred to as the Younger Dryas, was apparently global in scale and is usually attributed to a change in the circulation of the Atlantic Ocean. It is interesting to note that the final cold-to-warm transition that marked the end of the Younger Dryas appears to have taken place over a time interval as brief as 20 years, highlighting the fact that important changes in climate can take place extremely rapidly—something to bear in mind as we contemplate the changes in climate that may arise in the immediate future. Warming resumed subsequently, reaching a maximum between about 8,000 and 5,000 years BP (during the period referred to as the Hypsithermal). Temperatures have declined more or less steadily since, interrupted by modest warming between about AD 950 and 1250 (the Medieval Optimum) followed by an interval of much colder conditions (the Little Ice Age) between about AD 1500 and 1850. The longer record of past climates and the role imputed to changes in the Earth's orbital properties suggest that by the time of the Little Ice Age the Earth might have been well on its way to the next Big Ice Age. In this case, we humans, by adding large concentrations of CO_2 to the atmosphere by burning coal, oil, and natural gas, may have saved the day. But, as we will discuss later, by this point we may have gone too far!

So, to respond to those who would assert that the world has been both warmer and colder in the past, let me agree. I would take issue, however, with those who would argue that humans would have been able to accommodate to these changes. Remember, our species was not around during the warmer conditions that prevailed a few tens of millions of years ago and earlier. For as long as we have been around, concentrations of key greenhouse gases such as CO_2 have never been as high as they are today. That is a matter of fact not theory, borne out by measurements of gases from ancient atmospheres trapped in polar ice. Lessons from the past are of dubious value as a guide to the conditions we are likely to confront in the future.

ORGANIZATIONAL PLAN

This is a book primarily about energy. It is motivated by my sense of an overriding imperative that we must move as quickly as possible to address the challenge of unsustainable future climate change. I chose for a number of reasons to focus primarily on the United States. I elected at the same time to include discussion of the key elements of the situation in China, reflecting the fact that China is now the world's second largest economy (trailing only the United States), the world's largest consumer of coal, the second largest consumer of oil (trailing again only the United States), and the world's largest emitter of CO_2 (having surpassed the United States for this dubious distinction in 2006). There are several reasons to emphasize the United States. First, the United States is the world's largest economy, likely to enjoy this status for at least the next few decades. Second, if we are to make the transition successfully from the current fossil carbon-based global energy economy to a more climate-friendly alternative, I believe that leadership from the United States, the world's technologically most inventive society (from Edison and Ford to today's Internet, Facebook, Twitter, iPod, and iPhone), will be critical and consequential. The need to respond to the unusually fractious nature of the debate on climate change in the United States provides additional motivation. A vocal minority in the United States is disposed to reject out of hand even the possibility that humans could alter climate. In too many cases, the views expressed by those who take this position appear to be based on ideology rather than objective analysis—misplaced religious conviction, distrust of science, suspicion of government, or a feeling that those in authority are trying to use the issue to advance some other agenda, to increase taxes, for example. There is an urgent need for more informed discourse on the topic, for educators to educate and for leaders to lead. My hope is that this book can contribute to this objective.

Chapter 2 provides an introduction to energy basics: what we mean by energy, how we use it, what we pay for it, where it comes from, and how much we depend on it. Chapter 3 outlines how energy is deployed on a national scale in both the United States and China, including a summary of related emissions of CO_2. Chapters 4 and 5 take up the issue of human-induced climate change, presenting arguments in the former case as to why you should take it seriously, with contrary views discussed in the latter. Chapters 6–14 outline the status, problems, and prospects for individual energy choices, treating sequentially issues relating to coal, oil, natural gas, nuclear, wind, solar, and other possibilities, including hydro, geothermal, and biofuels. The historic agreement announced by US President Obama and Chinese President Xi in Beijing on November 12, 2014, to limit emissions of greenhouse gases from their joint countries over the near term is discussed in Chapter 15. The book concludes in Chapter 16 with the outline of a vision for a fossil carbon-free future energy system together with an account of the steps that must be taken if we are to achieve this objective, including comments on actions we may need to take should we fail to address this objective.

References

Fairbank, J. K., and M. Goldman. 2006. *China: A new history*. Cambridge, MA: Belknap Press of Harvard University Press.

Hoffman, P. E., A. J. Kaufman, G. P. Halverson, and D. P. Schrag. 1998. A Neoproterozoic Snowball Earth. *Science* 281: 1342–1346.

Imbrie, J., and K. P. Imbrie. 1986. *Ice Ages: Solving the mysteries*. Cambridge, MA: Harvard University Press.

Kirschvink, J. L. 1992. Late Proterozoic low-latitude global glaciation: The Snowball Earth. In *The Proterozoic biosphere*, ed. J. W. Schopf and C. Klein, 51–52. Cambridge, England: Cambridge University Press.

McElroy, M. B. 2002. *The atmospheric environment: Effects of human activity*. Princeton, NJ: Princeton University Press.

Perlin, J. A. 1989. *Forest journey: The role of wood in the development of civilization*. Cambridge, MA: Harvard University Press.

2

Energy Basics

IT IS IMPORTANT to have a basic understanding of what energy is, how we use it, how we measure it, what we pay for it, and where it comes from—the ultimate sources. You don't need a degree in physics to understand energy. You know intuitively what it is. You purchase energy when you buy gasoline or diesel oil to drive your car or truck. You use energy, most commonly in the form of either oil or natural gas, to heat your house in winter and to supply you with a source of hot water. The food you eat represents a source of energy. You may use natural gas to cook your meals. Or perhaps you do this using electricity, another essential form of energy.

It is easy to lose sight of the multiple ways in which we rely on electricity—for lighting; to run our radios, television sets, refrigerators, and freezers; to power our computers, cell phones, and elevators; to pump water; to wash our dishes and our clothes; and to run our air conditioning systems in summer—the list is almost endless. Just think, if you were to take a trip back in time, as little as 50 years or so, few of these electrical conveniences would have been available. Nor would you have needed a source of gasoline: the automobile was a play toy for the rich when it was first introduced in the last few years of the nineteenth century; it became popular only later, after 1908 when Henry Ford (1863–1947) introduced what became known as the people's car, the Model T.

Access to commercially available sources of energy allows us to carry out functions that would be otherwise difficult if not impossible. Exactly how should we define what we mean by energy? The language of science is very precise. The energy of a physical system is defined as the capacity of the system to do work. Work, in turn, is defined as the energy expended if a system is displaced a specified distance in opposition to an impressed force. Force is identified as the rate of change in the momentum of the physical system. Momentum is defined as the product of the mass of the system and its velocity. To add to the complexity, we also need a system of units with which to measure these quantities. Specifically we need units to define standards for the basic quantities—length (distance), mass, and time. The standard for scientists is Le Systeme Internationale, SI for short, which came into common practice in October 1960. In the SI system, length is measured in meters (m), mass in kilograms (kg), and time in seconds (s). All other quantities can be defined in terms of these basic units. The unit of energy in the SI system is the joule (J), named in honor of the English physicist James Joule (1818–1889). This is probably more than you need to know.

EVERYDAY FACTS

In practice we don't use joules to refer to the various forms of energy we employ in our everyday lives. We purchase gasoline and/or diesel oil in units of gallons to drive our cars and trucks, and similarly for the oil we use to heat our houses. A gallon, of course, is not a unit of energy but rather a unit of volume. Our bills for natural gas are quoted typically in units of therms. The unit in this case refers to a quantity of gas corresponding to an energy content of 10^5 (100,000) British thermal units (BTUs): 1 BTU defines the energy required to raise the temperature of a pound of water (454 kg) by one degree Fahrenheit. Our electricity bills are quoted in units of kilowatt hours (kWh). The watt (W), named in honor of James Watt (1736–1819), the Scottish inventor/scientist credited with perfection of the steam engine, is a unit of power rather than energy, the quantity of energy available per unit time: 1 W defines a supply of energy at a rate of 1 J per second. One kWh specifies the energy delivered when a power source of 1 kW is deployed for a period of 1 hour, equivalent to 3,600,000 joules. Run ten 100 W light bulbs for an hour and you consume 1 kWh of energy (1 kWh is equivalent to 1,000 watt hours (the prefix kilo means multiply by 1,000).

It is convenient to adopt a single unit when we think about our personal energy consumption. A typical human consumes energy in the form of food at an approximate rate of about 100 W (comparable to the power required to operate a 100 W light bulb). That means that each of us needs the equivalent of about 2.4 kWh per day to maintain our essential bodily functions (to breathe, to keep the blood flowing, etc.). The McElroy house in Cambridge, Massachusetts, is heated in winter with hot water supplied by a natural gas–fueled furnace. We use gas also for cooking and for hot water. Between August 2013 and August 2014, we consumed gas at an average rate of 6.2 therms per day, equivalent to 182 kWh per day (1 therm = 29.3 kWh), the bulk of this in winter during the heating season (the winter of 2013–2014 was unusually cold in New England and indeed across the United States). Our average daily demand for electricity over the same period amounted to about 15 kWh. There are two of us: this implies that our per-person daily consumption of natural gas and electricity added up to about 98 kWh, approximately 40 times the content of energy in the food we eat to stay alive.

Expressed in kWh, the energy content of a gallon of gasoline is equal to 32.63 kWh. The average passenger car in the United States is driven approximately 12,000 miles per year with an average fuel efficiency of about 22.5 miles per gallon, corresponding to a daily demand for fuel of 1.46 gallons, in turn corresponding to an energy requirement of 47.6 kWh. My wife and I drive significantly less than the national average and our cars are more efficient than the average. I estimate our per capita daily average energy demand for gasoline at 21 kWh. Adding it up, excluding the food we eat, our daily per capita energy total rises to about of 119 kWh, and the multiplier with respect to minimum requirements for survival expands to about 50.

We should probably add to our energy bill, though, the energy for which we are responsible when we travel by plane. David MacKay[1] points out that if you take as little as one intercontinental plane trip per year, your personal energy budget would rise on average by as much as 30 kWh per day. A more accurate estimate of our extended personal daily energy budget is probably closer to 170 kWh per person. Do your own calculation (useful website noted in the next section).

WHAT WE PAY FOR ENERGY

Retail prices for gasoline, natural gas, and electricity in Cambridge, Massachusetts, averaged approximately $3.50 per gallon, $1.75 per therm, and 19.3 cents per kWh, respectively, in 2013–2014. The data quoted here for natural gas and electricity refer to prices paid for delivery to residential consumers: prices for large industrial and commercial customers are typically somewhat lower, reflecting more predicable patterns in demand and generally lower expenses for delivery. Converting these data to costs per kWh of energy, Cambridge residents paid 10.8 cents per kWh for gasoline, 6.1 cents per kWh for natural gas, and, as indicated earlier, 19.3 cents per kWh for electricity.

On a per unit energy basis, electricity is expensive. It is easy to understand why that should be the case. Electricity is classified as a secondary energy source. It must be produced from what is identified as a primary source. On a national basis, 39% of US electricity in 2013 was generated using coal, 27% using natural gas, 19% from nuclear, 7% from hydro, and 4% from wind, with the balance from a combination of oil (1.0%), wood (1.0%), other biomass (0.5%), geothermal (0.4%), and solar (0.2%). The relatively high price of electricity on an energy basis compared to coal, natural gas, or oil reflects the relatively low efficiency with which these energy sources are converted to electricity, typically less than 40%. Efficiencies are particularly low for some of the nation's older coal-fired plants, as low as 20%, though generally better for natural gas–fired plants, especially for modern gas-combined-cycle systems for which efficiencies can range as high as 50%.[2] The fraction of the fuel energy that is not converted to electricity is released to the environment in the form of waste heat (heat is also a form of energy).

Prices for electricity vary significantly across the United States, as indicated in Figure 2.1 (August 2014 data). Prices range from a high of 37.8 cents per kWh in Hawaii

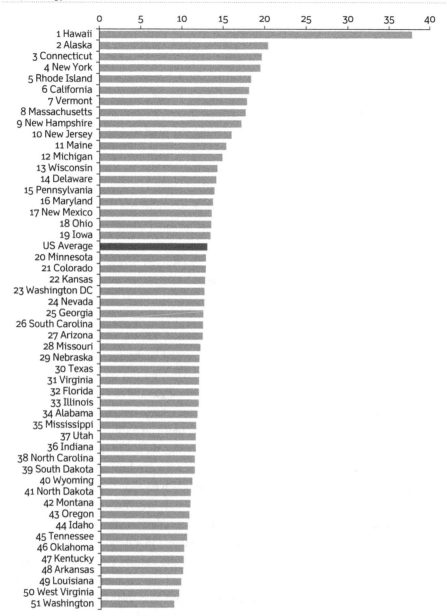

FIGURE 2.1 State ranking for residential electricity prices for August 2014. (*Data source:* http://www.eia.gov/state/rankings/?sid=US#/series/31)

(clearly an outlier) to a low of 8.93 cents per kWh in Washington with a national average of 13.0 cents per kWh. Four of the 10 states with the lowest prices for electricity (Kentucky, West Virginia, Indiana, and Montana) generate the bulk of their power using relatively inexpensive, locally produced coal. Hydropower, developed using dams built and paid for many years ago, provides the dominant source of electricity for three of the states in the bottom 10 (Idaho, Washington, and Oregon). Seven states in the

Northeast (Connecticut, New York, New Jersey, Rhode Island, New Hampshire, Vermont, and Maine), joined by California, rate in the top 9 in terms of price. What these states have in common is a willingness to pay a premium for power to minimize emissions of pollutants that could adversely affect their air quality and the overall health of their environments. They use little coal, the most polluting of the fuels, in their generation mix. The bulk of their electric power is derived from a combination of natural gas, nuclear, and hydro with important contributions from (expensive) petroleum in Connecticut and New York.

My colleague Dr. Xi Lu drew my attention to an interesting website (http://www.epa.gov/cleanenergy/energy-and-you/how-clean.html) that can let you find out how electricity is produced in your local region, together with the related emissions, simply by specifying your zip code and your electricity provider. I recommend this site to readers who may wish to explore their personal energy budgets in more detail.

Prices of electricity in the United States are largely independent of conditions in the external world environment. As indicated earlier, 66% of US electricity is generated using domestically produced coal and natural gas with nuclear sources accounting for an additional 19%. The United States has abundant domestic supplies of coal, natural gas, and uranium to fuel its existing and anticipated future coal, natural gas, and nuclear power plants. The situation is very different, however, for the energy sources deployed in the transportation sector—gasoline, diesel oil, and aviation fuel—where prices are unusually volatile, responding to subtle changes in the international balance between supply and demand.

The price of oil hit a peak of $147 a barrel (a barrel corresponds to 42 gallons) in July 2008 in advance of the financial crisis, falling to a low of $37.58 a barrel on December 26, 2008, before beginning a slow recovery. It had returned to about $100 a barrel by the beginning of 2011 and hovered around this level until the end of September 2014 before again beginning to drop. This most recent decline caught the financial markets by surprise. Experts had predicted that the price of crude oil was more likely to rise rather than fall given disruptive events in oil-producing regions of the Middle East, notably the rise of the terrorist Islamic State in Iraq and Syria organization (ISIS). By December 14, 2014, the price for January 2015 futures of West Texas Intermediate crude (WTI), the benchmark for US oil prices, had dropped below $58 a barrel and gasoline prices in the United States had fallen below $3 a gallon for the first time since early 2011. The future trajectory of oil prices is essentially unpredictable (I am writing this in late December 2014). What is clear is that increased extraction and reduced consumption of oil in the United States (discussed in greater detail in Chapter 7) have had a major impact on prices for crude oil and oil products not only in the United States but also globally. The relatively sluggish state of the global economy has also had an impact. Underscoring the significance of the recent changes in the US oil economy, net imports in the United States as a share of overall demand for liquid fuels decreased from 60% in 2005 to 33% in 2013. The United States is now a net exporter of petroleum products.

And China, as of September 2013, has replaced the United States as the world's largest net importer of crude oil and oil products.

Prices for gasoline in the United States are cheap compared with prices in either Europe or Japan. Costs for a gallon of gasoline (US measure as distinct from the larger imperial version) averaged, respectively, $9.10, $7.95, $8.28, and $8.50 in the Netherlands, France, Germany, and the United Kingdom in July 2014. Motorists in Japan paid $5.26/gallon. The price in mid 2015 in the United States averaged less than $2.80/gallon. The base price for gasoline should be comparable for all countries, determined mainly by the international market price of oil. Differences across countries reflect therefore differences in taxes. The federal tax for gasoline in the United States amounts to a modest 18.4 cents a gallon and has remained at this level since 1993. *New York Times* columnist Tom Friedman has suggested that $4-a-gallon gasoline represents "a red line where people really start to change their behavior." He proposed that "the smart thing for us to do right now is to impose a $1-a-gallon gasoline tax, to be phased in at five cents a month beginning in 2012, with all of the money going to pay down the deficit." Think what that would have done to our fiscal deficit—additional income to the federal treasury of $138 billion per year (assuming current levels of consumption). And it would still have left us with gasoline much cheaper than for most of the countries we consider our economic competitors. Chances for passage of such a tax in the United States are slim, however. Access to cheap gasoline in the United States is considered just a little less than a guaranteed birthright and woe befall the politician who would tinker with it.

A summary of net US gasoline taxes on a state-by-state basis is presented in Figure 2.2. Net taxes are highest in New York and, not surprisingly, lowest in Alaska, where individual Alaska-born citizens continue to receive annual checks from the Permanent Fund set up in the state to administer income from the state's oil production. The payout in 2014 amounted to $1,884 per person, a modest drop from the record payout of $2,069 in 2008.

ULTIMATE SOURCES

The energy of coal, oil, natural gas, wood, and biomass is stored primarily in the form of the chemical bonds that bind carbon atoms to carbon atoms and carbon atoms to hydrogen atoms. Natural gas, for example, is composed mainly of methane in which individual carbon atoms are linked to four hydrogen atoms, forming a molecular structure identified as CH_4. When the gas is combusted (burned), its chemical form is altered. The carbon and hydrogen atoms link up with oxygen atoms supplied from the air that fuels the combustion process. If combustion proceeds to completion, the carbon of the methane is converted to carbon dioxide (CO_2), and at the same time the hydrogen is converted to water (H_2O). Energy is released in the process: the energy content of methane is significantly higher than the combined energies of the CO_2 and H_2O molecules

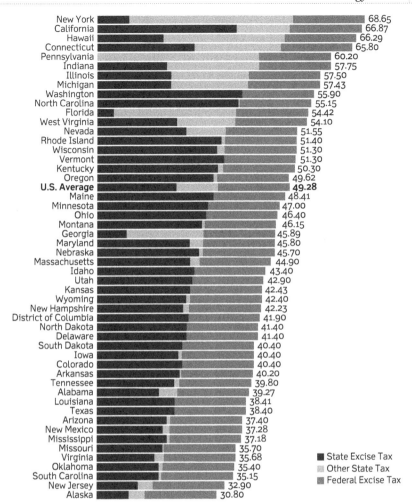

FIGURE 2.2 Summary of net US gasoline taxes (cents per gallon) on a state-by-state basis in October 2014.
(*Data source:* http://www.api.org/oil-and-natural-gas-overview/industry-economics/~/media/Files/Statistics/State-Motor-Fuel-Taxes-Report-October-2014.pdf)

produced as a result of the combustion process. The extra energy is realized in the form of heat (total energy is conserved). Carbon in the form of coal, oil, natural gas, wood, and biomass is said to be chemically reduced (think high energy). Carbon in the form of CO_2 is oxidized (low energy, useless as a fuel). Coal, oil, natural gas, and biomass represent sources of what we refer to as chemical potential energy (energy that is stored chemically but can be released by combustion).

What is the origin of the chemical energy stored in these fuels? The proximate source is the sun. Sunlight captured by the chlorophyll of green plants (including phytoplankton in the ocean and land-based aquatic systems) was used to convert CO_2 and H_2O to the higher energy (reduced) forms of carbon that provide eventually not only the energy source for the plants but also, after some chemical rearrangements, their structural material. The process is known as photosynthesis. All of life, from the largest animals to the smallest microbes, depends on this energy extracted from the sun. In particular, the sun is responsible for the energy we consume in the form of food, in addition to the energy that is used to feed the cattle, sheep, fish, and other intermediaries those of us who are not strict vegetarians exploit to supplement our diets.

The chemical potential energy of coal, oil, and natural gas is a product of sunlight absorbed by plants that lived on the Earth (both on land and in the sea) several hundred million years ago. This plant material was transformed over time by a variety of biological and geological processes to form the fuels we harvest today.[3] For obvious reasons, we refer to coal, oil, and natural gas as fossil fuels. They provide a valuable means to store a portion of the energy from the sun that fell on the Earth in the distant past. The fossil source of energy is, of course, limited and exploiting it, as we will discuss later, is not without its problems. In particular, when we burn fossil fuels, we are returning carbon to the atmosphere that has been absent from the atmosphere, buried in sediments, for several hundred million years.[4] This fossil carbon is released in the form of CO_2, a potent greenhouse gas (more on this later when we talk about the challenge of human-induced climate change). Given the large quantities of fossil fuels we humans have consumed over the past few hundred years, it is surely not surprising that the abundance of CO_2 is now at a record high. In fact, it is higher, as we shall see, than at any time over the past 800,000 years, potentially greater than at any time since dinosaurs roamed the Earth 50 million years ago.

The Department of Energy refers to coal, oil, natural gas, wood, biomass, nuclear, hydro, wind, solar, and geothermal as primary sources of energy with electricity identified as secondary. This is accurate in the sense that electricity is produced from a combination of all of these resources. It is inaccurate, however, in that the primary sources we refer to are themselves derivative. This is clear from the foregoing discussion with respect to the fossil fuels, wood, and biomass. Solar energy is responsible also for the energy sources represented by hydro and wind. The sequence of transformations involved with hydro is as follows: sunlight absorbed primarily by the ocean results in evaporation of water; a large amount of energy is consumed in changing the phase of water from liquid to vapor; vapor condenses in the atmosphere to form either liquid or ice, converting the energy consumed in evaporation to heat in the atmosphere; a portion of the condensed phase precipitates, it falls to the surface as either rain or snow; if the precipitation reaches the surface at an elevated altitude, it represents a source of potential energy, that is to say, the water can pick up speed as it moves to lower elevations (acquire what we refer to as kinetic energy, the energy of motion); alternatively, the precipitation can be collected and stored in dams, building up pressure in proportion to the depth of stored water; this pressure can be used to drive water through a turbine (converting potential to kinetic energy); the turbine transforms a portion of this energy to electricity. The sequence for wind is similarly complex. There are two ways to generate kinetic energy in the atmosphere. It can be produced if air descends, sinks from a higher to a lower altitude. Or it can be formed if air moves from a region of high pressure to one of lower pressure. The force that acts on the air to do work in the former case is the force of gravity. In the latter case it is what is referred to as the pressure gradient. Electricity is generated from wind when the kinetic energy of wind is captured and used to turn the blades of a wind turbine.

With the exception of a minor contribution from heat originating in the Earth's interior,[5] the sun, as we have seen, is the dominant source of the energy we exploit to fuel our relatively comfortable modern lifestyles and indeed to maintain the global climate system and the diversity of life on our planet. This energy arrives at the Earth's surface largely in the form of visible light (radiation) emitted from a region of the solar atmosphere known as the photosphere, where the temperature is close to 5,600°C. The energy the sun emits as light is produced by nuclear reactions in the high-pressure, high-temperature core of the sun, in a region where pressures are as much as 300 billion times higher than the pressure at the surface of the Earth and where temperatures reach levels as high as 15 million°C. In this environment, protons (the nuclei of hydrogen atoms) are squeezed so closely together that they can combine to form heavier nuclei.[6] We refer to this process as nuclear fusion. In the process, mass is converted to energy consistent with Einstein's famous statement of the equivalence of mass and energy: $E = mc^2$, where E denotes energy, m is mass, and c is the velocity of light. In this sense, nuclear reactions represent the ultimate primary source of energy for our planet.[7]

Notes

1. David MacKay's book, *Sustainable Energy—Without the Hot Air*, as indicated in its preface, is devoted to "cutting UK emissions of twaddle-twaddle about sustainable energy." The book does an excellent job in identifying how energy is used on a per capita basis in the United Kingdom and the choices that must be made if the United Kingdom is to reduce its emissions of greenhouse gases. It was MacKay's lead that persuaded me to express personal energy use in terms of kWh/person/day. The present book is directed more at a US audience. MacKay makes the point in Chapter 1 that "in a climate where people don't understand numbers, newspapers, campaigners, companies and politicians can get away with murder." He may have directed this criticism at media, businesses, and politicians in the United Kingdom, but it would appear that his critique applies at least equally to constituencies in the United States.

2. In conventional coal- and gas-fired steam power plants, energy released when the fuel is combusted is used to produce high-pressure steam. This steam is allowed to expand, turning the blades of a turbine to generate electricity. When the useful energy of the steam is depleted, the residual steam is cooled, normally by contact with water drawn from some convenient source such as a river, a lake, or the ocean. The resulting condensed liquid is pumped back to the boiler, completing the cycle. Production of electricity using a conventional steam-driven power plant requires a significant source of water as coolant. This is true not only for coal and gas-fired steam plants but also for nuclear plants. In 2003, during an unusually hot summer, the French nuclear industry was obliged to shut down a number of its nuclear plants when the temperature of waters

in the rivers normally used to provide coolant rose to a level such that when these waters were used to extract energy from the depleted steam, the temperature at which the water would have been returned to the rivers was judged to be unacceptably high. Natural-gas-combined-cycle (NGCC) plants are more efficient than conventional coal or gas steam systems. Electricity is produced in NGCC systems using energy extracted not only from the hot gases produced in the initial combustion process but also by capturing a portion of the heat left over from this process to produce steam that is used to drive a second turbine. The more efficient use of chemical energy in the NGCC process accounts for its greater efficiency, as high as 50% as indicated in the text.

3. Most of the world's coal was formed during the geological era known as the Carboniferous Period approximately 300 million years ago. The plants that captured the solar energy that resulted eventually in coal grew for the most part in tidal and/or deltaic environments in the tropics and subtropics. The landmasses of the Earth were arranged very differently in the Carboniferous Period as compared to present, a circumstance that accounts for the current concentration of major coal deposits in North America, Europe, China, and Indonesia. The plant progenitors of coal grew in swampy environments. When the plants died, the material of which they were composed fell into the waters of the swamps, where insufficient supplies of oxygen inhibited their decomposition. The plant material was buried subsequently, covered by sediments derived from erosion of continental rocks during a period when climate oscillated between warm interglacial and cold glacial conditions. The first step in coal development involved deposition of a relatively thick residue of plant material forming peat. Subjected to an increasing overburden of sediment, the peat was compacted. Gradually it was transformed to low-grade brown coal or lignite. With exposure to increasingly higher temperatures and pressures, the lignite was converted to progressively higher energy forms of coal, initially sub-bituminous, subsequently bituminous, and eventually the highest grade of all, anthracite. In contrast with the case of coal, where the parent plant material grew mainly in terrestrial environments, the organisms responsible for oil lived for the most part in the ocean. A necessary condition for the production of an economically viable concentration of oil was a locally high rate of biological productivity, prompted most likely by an increased source of nutrients, specifically nitrogen and phosphate, supplied either from the land (in estuaries, for example) or by upwelling from the deep (in coastal zones, for example). A second condition was that the rate at which organic matter was added to sediments should exceed the rate at which oxygen was supplied from the overlying water (that would otherwise have supported a population of oxygen-breathing organisms capable of effectively consuming the incoming organic matter). As the organic matter was buried, it was subjected to increasingly higher temperatures. Temperatures in the range of 60°C to about 160°C, realized typically at depths of between 2 and 6 km, were required to produce the chemical building blocks of oil. The final requirement was that the oil should be preserved, trapped beneath a layer of impermeable rock (shale, for example). At higher temperatures and pressures (depths greater than 7 km), the organic matter would have been broken down to form low molecular weight compounds such as methane (CH_4), ethane (C_2H_6), and butane (C_4H_{10}), contributing to the production of what we identify as natural gas. As with oil, survival of natural gas for eventual exploitation depends on a suitable geological regime in which the gas can be trapped. Natural gas is also commonly produced in conjunction with the biological and geological processes responsible for production of coal.

4. The bulk of the Earth's carbon is stored in sediments. The carbon in the combined atmosphere-biosphere-ocean system has a residence time of a little less than 200,000 years, in

contrast to the residence time of hundreds of millions of years for carbon in sediments. The carbon in sediments is eventually cycled back to the atmosphere, thanks to the influence of plate tectonics. The sediments are either uplifted and weathered or drawn down into the mantle, where their constituent carbon can be cooked and returned to the atmosphere, in the latter case as a component of either hot springs or volcanoes. As discussed in Chapter 1, changes in sedimentary cycling rates have resulted over time in a variety of different levels of carbon in the atmosphere-biosphere-ocean system, contributing to a range of different climate systems. Mining coal and drilling for oil and gas may be considered as a human-assisted acceleration of the rate at which carbon is returned to the atmosphere from sediments. Our contribution at present exceeds that of nature by more than a factor of 50. It is not surprising under the circumstances that the concentration of CO_2 in the atmosphere is increasingly accordingly.

5. Temperatures increase by about 20°C to 30°C per kilometer as a function of depth below the Earth's surface. Decay of radioactive elements such as uranium, thorium, and potassium is largely responsible for production of the energy responsible for this rise in temperature. As discussed later in Chapter 13, drilling into the Earth could provide an opportunity to tap this source of energy. Cold water injected into a deep well could be heated by exposure to hot rock and converted to steam. The steam could return to the surface in another well, where it could be used to drive a turbine and produce electricity. Informed opinion (discussed later in Chapter 13) suggests that geothermal energy could make an important, economically viable, sustainable contribution to our demand for electricity in the future.

6. Protons (the nuclei of hydrogen atoms) are squeezed so close together in the sun's core that they combine initially to form deuterium (nucleus composed of two protons). Subsequent reactions result in the conversion of four protons to an alpha particle, the nucleus of a helium atom, and eventually to even heavier nuclei. Conversion of mass to energy consistent with Einstein's relation is ultimately responsible for the vast quantities of electromagnetic radiation (light) emitted by the sun. A pipe dream for energy scientists is to reproduce the conditions for fusion of hydrogen on Earth, to develop a commercially viable system that could provide an essentially unlimited source of nonpolluting energy. While there has been limited success in demonstrating aspects of the fusion opportunity in laboratory experiments, prospects for development of a viable commercial scale application lie at least 20 years in the future. But, to add a morsel of realism, predictions for an economically viable future for fusion have pointed to 20 years in the future for at least the past 50 years!

7. As indicated here, nuclear reactions account for the bulk of the energy ultimately available on the Earth. The single exception perhaps is the energy associated with tides. The energy in this case is extracted from work performed by the gravitational forces that bind the Earth to the sun and moon. Particularly important is the influence of the lunar gravitational field in raising the surface of the ocean in the regions closest to and furthest from the moon (the rise in sea surface in the latter case results from the force exerted on the solid Earth, which experiences a pull toward the moon, allowing water to pile up in the space vacated as a consequence). Tidal energy is dissipated by friction between ocean water and the underlying surface. Conversion of gravitational energy to heat at the Earth's surface has resulted over geological time in a slow increase in the length of the day (a decrease in the rotation rate of the Earth) and a slow increase in the distance separating the Earth from the moon. The changes in both cases are modest, detectable only with extremely sensitive instrumentation. As discussed in more detail in Chapter 12, water as it ebbs and flows twice daily

could be used to drive a turbine and produce electricity. The first tidal energy system was installed at La Rance in France in the early 1960s. Small-scale systems are currently operational in Strangford Lough in Northern Ireland and in China (the Jianxia Tidal Station near Hangzhua), and plans are in place to develop additional systems elsewhere. Despite its potential for localized regions where tidal amplitudes are particularly large, tidal energy is unlikely to make a significant contribution to the world's future demand for energy either locally or globally. Costs in general are too high, and there are problems relating to environmental disruption in regions where these systems might be deployed.

3

The Contemporary US Energy System

OVERVIEW INCLUDING A COMPARISON WITH CHINA

THE DISCUSSION IN Chapter 2 addressed what might be described as a microview of the US energy economy—how we use energy as individuals, how we measure our personal consumption, and how we pay for it. We turn attention now to a more expansive perspective—the use of energy on a national scale, including a discussion of associated economic benefits and costs. We focus specifically on implications for emissions of the greenhouse gas CO_2. If we are to take the issue of human-induced climate change seriously—and I do—we will be obliged to adjust our energy system markedly to reduce emissions of this gas, the most important agent for human-induced climate change. And we will need to do it sooner rather than later. This chapter will underscore the magnitude of the challenge we face if we are to successfully chart the course to a more sustainable climate-energy future. We turn later to strategies that might accelerate our progress toward this objective.

We elected in this volume to focus on the present and potential future of the energy economy of the United States. It is important to recognize that the fate of the global climate system will depend not just on what happens in the United States but also to an increasing extent on what comes to pass in other large industrial economies. China surpassed the United States as the largest national emitter of CO_2 in 2006. The United States and China together were responsible in 2012 for more than 42% of total global emissions. Add Russia, India, Japan, Germany, Canada, United Kingdom, South Korea, and Iran to the mix (the other members of the top 10 emitting countries ordered in terms of their relative contributions), and we can account for more than 60% of the global total.[1] Given the importance of China to the global CO_2 economy (more than 26% of the present global total and likely to increase significantly in the near term), I decided that it would be instructive to include here at least some discussion of the situation in China—to elaborate what the energy economies of China and the United States have in common, outlining at the same time the factors and challenges that set them apart.

US ENERGY USE AND EMISSIONS

Figure 3.1 provides a summary of how energy was used in the United States in 2013. Related emissions of CO_2 are displayed in Figure 3.2. Temporal trends in energy use and CO_2 emissions are displayed in Figures 3.3 and 3.4, respectively. The energy data are presented here in units of quads: 1 quad is equal to a quadrillion (10^{15}) BTU or 1,000 trillion BTU.[2] Emission data for CO_2 are quoted in terms of millions of tons of CO_2.[3] The downturns in energy use and CO_2 emissions beginning in 2009 were due in part to the recession that started in late 2008, in part because of a price-driven substitution of natural gas for coal in the production of electricity resulting from a significant increase in the supply of gas from shale (a topic to be discussed in further detail in Chapter 8). The structure exhibited by the energy consumption curve (Fig. 3.3) during the 1970s reflects separately the influences of the Yom Kippur War (1973) and the Iranian hostage crisis (1978). Supplies of oil were disrupted on both occasions. Prices rose by more than a factor of 3 over periods as brief as a few months (by as much as a factor of 9 over the decade), driving the economy in both cases into serious recession.

The key features of the US energy economy as summarized in Figure 3.1 are as follows:

1. Generation of electricity accounted for a little more than 39% of total primary energy use in the United States in 2013.
2. Coal was responsible for 39% of US electricity production in 2013, followed by natural gas (28%), nuclear (19%), and hydro (7%), with the balance from a combination of wind (4.2%), biomass (1.5%), oil (1%), geothermal (0.4%), and solar (0.2%).
3. Sixty-eight percent of the energy deployed to generate electricity provided no useful function—it was rejected to the environment as waste heat.
4. Petroleum accounted for 36% of total US consumption of primary energy in 2013, the bulk of which (71%) was deployed in the transportation sector.
5. Seventy-nine percent of the energy used in the transportation sector failed to perform any useful function—it was lost as waste heat.
6. Transportation accounted for the largest fraction of final energy use in the United States in 2013 (37.7%), followed by contributions from the industrial (34.4%), residential (15.9%), and commercial sectors (12.0%).

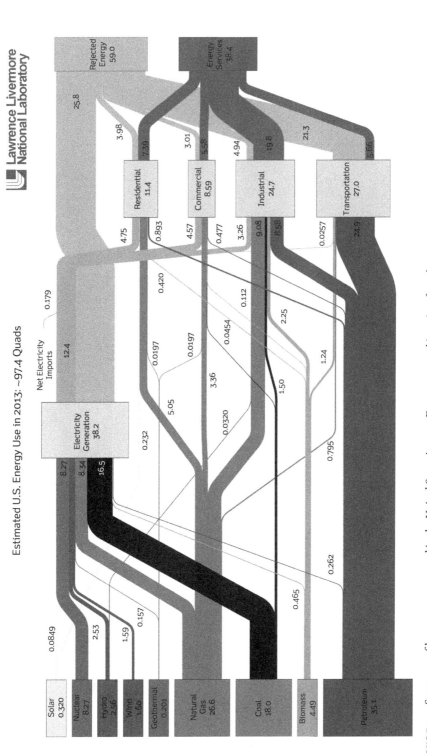

FIGURE 3.1 Summary of how energy was used in the United States in 2013. Data are quoted in units of quads.
(*Source*: https://www.llnl.gov/news/newsreleases/2014/Apr/images/31438_2013energy_high_res.jpg)

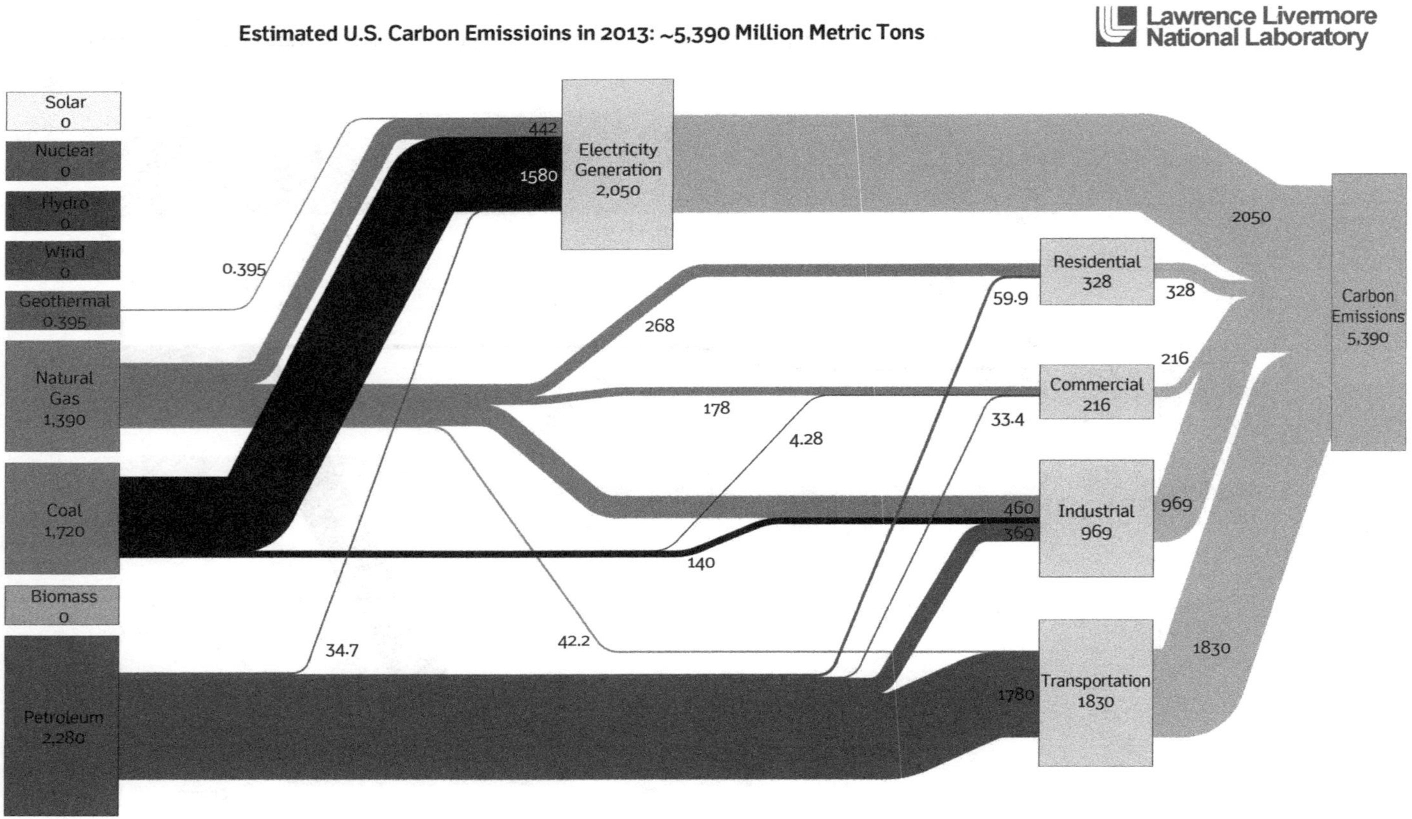

FIGURE 3.2 Emissions of CO_2 associated with energy consumption in the United States in 2013. (*Source:* http://blogs.app.com/enviroguy/files/2014/04/April46.png)

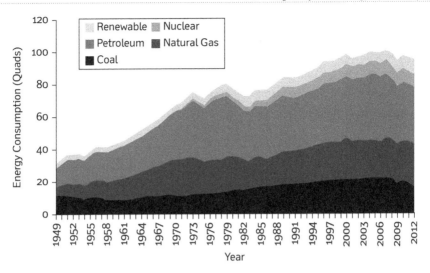

FIGURE 3.3 Temporal trends for annual energy use in the United States.
(*Source:* http://www.eia.gov/totalenergy/data/annual/showtext.cfm?t=ptb0102)

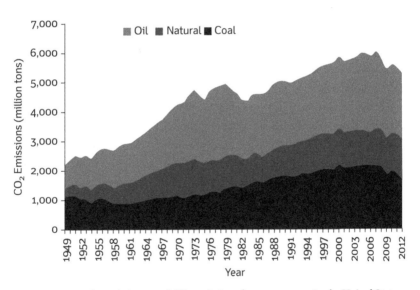

FIGURE 3.4 Temporal trends in annual CO_2 emissions from energy use in the United States.
(*Source:* http://www.eia.gov/totalenergy/data/annual/showtext.cfm?t=ptb1101)

The key features of the CO_2 emission data in Figure 3.2 are that:

1. Combustion of coal associated with production of electricity was responsible for 29% of total US CO_2 emissions in 2013.
2. Allowing for the contributions from natural gas and oil, the electricity sector accounted for 38% of total US CO_2 emissions in 2013.

3. Thirty-four percent of US emissions were derived from the transportation sector in 2013, comparable to the contribution from the generation of electricity.
4. The industrial, residential, and commercial sectors accounted, respectively, for 18%, 6%, and 4% of total US emissions in 2013, shares that should be adjusted upward to allow for consumption of electricity in these sectors, contributions from which are not explicitly recognized in Figure 3.2.

CHINESE ENERGY USE AND EMISSIONS

The structure of the Chinese energy economy differs significantly from that of the United States. The economy of China is classified as developing. What that means is that important resources are devoted to establishing the infrastructure required for further growth—construction of roads, buildings, railroads, airports, and so forth, all of which require major expenditures of energy with limited immediate benefits to the national gross domestic product (GDP). Reflecting these circumstances, China was responsible for 58.6% of total global cement production and 48.5% of total global steel production in 2013. Coal accounted for 70.2% of total primary energy in China in 2007 followed by petroleum (19.7%), hydro (5.8%), and natural gas (3.4%) with minor contributions from nuclear, solar, and wind: 43% of primary energy was used to generate electricity, 89% of which was produced using coal. Consumption of primary energy increased in China by 53.8% between 2007 and 2013, from 74.6 quad to 114.8 quad. Coal continued to provide the dominant source (66.5% of the total), rising from 52.4 quad in 2007 to 76.3 quad in 2013, with oil, hydro, natural gas, and nuclear responsible, respectively, for 17.5%, 7.1%, 5.0%, and 0.9% of the 2013 total. A notable change between 2007 and 2013 was the emergence of a significant contribution to electric power production from wind, which accounted for 1.0% of the total energy use in 2013. Imports were responsible for 56.2% of China's total petroleum supply in 2013. In contrast to the United States, where transport accounts for the bulk of oil use, only about 41% of petroleum is deployed in the transportation sector in China; the balance is used as feedstock in a variety of industrial, commercial, and residential applications. Demand for oil is expected to increase significantly in China over the next few years, tracking the anticipated rise in the number of privately owned automobiles as the structure of the Chinese economy evolves from its current state of development to one more closely approximating that of the United States and other developed countries.

Reflecting the elevated energy intensity of the Chinese economy, 25,775 BTU was required to produce a dollar of GDP in China in 2007 (expressed in 2005 US dollar equivalents) as compared to 7,672 BTU to achieve a similar result in the United States. The energy intensity decreased marginally between 2007 and 2011, from 25,775 BTU to 24,725 BTU per dollar of GDP. In contrast to the high energy intensity of the overall economy, energy consumption per person in China in 2013 was approximately 3.4 times less than the comparable level per person for the same year in the United States (82.6 million BTU as compared to 285.1 million BTU), attesting to the greater wealth and higher levels of consumption for the American populace as a whole. China's earlier stated goal, responding to the threat of global climate change, was to reduce the carbon intensity of its economy by 40%–45% by 2020 relative to 2005: what that would mean is that China would be committed to emit 45% less CO_2 per unit of GDP in 2020 as compared to 2005. But, if China's economy were to continue to grow at a 10% per year rate as it has in the past (7% more recently), this would mean that China's emissions of CO_2 would increase by a factor of 2.5 by 2020 relative to 2005. Think of what that would imply: if the global climate system had to accommodate 1 China in 2005, by 2009 it had to deal with 1.4 Chinas (emissions grew by a factor of 1.4 over this period), and by 2020 it could have to accommodate to as many as 2.5 Chinas!

President Obama enunciated a goal for the US economy to reduce US emissions of CO_2 by 80% by 2050 relative to 2005. There is, of course, no assurance that this goal will be met given the contentious state of the current US political system with respect to the importance of the climate challenge. But, if it were, it would imply that the carbon intensity of the US economy would decrease from 0.48 kg CO_2 per dollar of GDP in 2005 to 0.04 kg per dollar of GDP in 2050, assuming annual growth of 1.5% for GDP over this period. Under the Chinese plan, the carbon intensity of the Chinese economy would decline from 1.02 kg/$ in 2005 to 0.61 kg/$ in 2020 (all monetary data here are quoted in terms of 2005 US dollar equivalents).

CONCLUDING REMARKS

The key conclusion to be drawn from this discussion is that if we are to markedly reduce US emissions of CO_2, we will need to find an alternative to coal as the primary source of energy for production of electricity and a substitute for petroleum-based products—gasoline, diesel, and jet fuel—as the energy source of preference for driving our cars, trucks, buses, and trains and for fueling our planes. Were we to eliminate coal and oil entirely from the nation's energy mix, we could reduce US emissions of CO_2 by as much as 78%. To accomplish this objective at the limit may be impossible, or at least unrealistic. But we can at least move in the direction of doing so, by reducing our dependence on coal and oil while at the same time preserving our economy and the quality of life we value and take for granted. China's challenge is even more daunting. China, as was the case until recently for the United States, is dangerously dependent on imported oil. Independent of the climate issue, there is an urgent need to address this dependence, which seriously threatens China's long-term security. Further, in the not too distant future, China will need to find an alternative to domestic coal as the dominant source of energy to fuel its economy. As we shall discuss later, in Chapter 6, given current patterns of coal use—growth in demand at an annual rate in excess of 10%—China is projected to exhaust its present proven domestic reserves in less than 30 years (China is now for the first time a net importer of coal). The United States, in contrast, is coal-rich: current US reserves are sufficient to accommodate domestic demand at present rates of consumption for more than 200 years. Is the United States destined to become a significant supplier of coal to China in the near future? What would that imply for the global climate system?

The success of civilizations in the past depended to a large extent on access to affordable sources of energy, combined with favorable local climates. Wood, for much of human history, accounted for the dominant primary source of energy. When wood ran out, or when climate underwent a significant change, civilizations collapsed.[4] As recent as 1900, wood was the most important source of energy for the United States. It has been supplanted since by fossil sources, by coal, oil, and natural gas, and similar transitions have occurred across the world.[5] The success of individual economies over the past hundred years can be attributed in no small measure to access to reliable, and for the most part inexpensive, sources of fossil fuels. But today, as discussed earlier, we face twin threats—uncertain supplies and the danger of potentially irreversible changes to the global environment (extinction of species and significant increases in global sea level, for example). We need no less than a new energy paradigm to ensure a harmonious future for the human enterprise. Should we fail to address this challenge we may well be destined to repeat the unfortunate experiences that led many years ago to the demise of civilizations as diverse as Sumeria, Egypt, Babylon, Crete, Greece, Cyprus, and Rome. As George Santayana famously remarked, those who forget their history are fated to repeat the mistakes of the past. I am confident, though, as elaborated later in this volume, that with proper foresight we can chart a more acceptable path to a more sustainable future.

Notes

1. Global action to deal with the challenge of human-induced climate change is organized under an agreement negotiated in Rio de Janeiro, Brazil, in June 1992. The treaty formalizing this agreement is known as the United Nations Framework Convention on Climate Change (UNFCCC). There are 194 national Parties to the Treaty. The US Senate ratified the Treaty in October 1992 (submitted by then President George H. W. Bush), joined a few months later (in January 1993) by China. The UNFCCC process is informed by scientific, technical, and socioeconomic information developed on a regular basis by the Intergovernmental Panel on Climate Change (IPCC). Reaching agreement among Parties with such a diverse range of interests is obviously difficult. We should point out, though, that if the top 10 countries noted here, joined by the remaining countries in the European Union, were to come to agreement on a common strategy to limit emissions, we could address in one fell swoop as much as 75% of the total global source. This could represent an effective alternative, or at least a useful supplement, to the cumbersome, often ineffective, meetings, referred to as the Conferences of the Parties (COPs), convened on an annual basis by the 194 current Parties to the UNFCCC.

2. National energy data are reported typically in units of quads. In the preceding chapter, I elected to quote personal energy use in units of kWh per person per day. The population of the United States in 2010 was equal to a little less than 309 million. Assuming an annual consumption of 100 quad of energy by 309 million people and converting to kWh, per capita daily US energy use would be equal to 260 kWh. In Chapter 2, I estimated that the McElroy family used

energy at a per capita rate of about 170 kWh per day. The comparison with the national data suggests that our family is actually doing quite well. But this is somewhat misleading since the national data reflect total use of energy, the combination of energy used to generate electricity, in addition to the energy consumed directly by individuals, commerce, and industry.

3. The data quoted here for annual US emissions of CO_2 imply that each of the 309 million residents of the United States is responsible for emission of approximately 19 tons of CO_2 per year. The CO_2 we emit is, of course, invisible, odorless, and nontoxic. It may come as a surprise that CO_2 is arguably the greatest (by either volume or mass) waste product for which we are personally responsible as a global society.

4. My earlier book (*Energy: Perspectives, Problems and Prospects*, Oxford University Press, 2010) includes an extensive chapter on the history of energy use going back to the time when our ancestors lived by hunting and gathering. The chapter includes an account of the role that changes in regional climate played in the demise of a number of important civilizations in the past. These changes were for the most part natural in origin: climate varies for a variety of reasons on a variety of time scales. The changes we are concerned about today are different. We humans are now a significant agent for this change, a force to be reckoned with comparable to the changes in the output of energy from the sun, in the orbital parameters of the Earth, in outgassing from the Earth's interior, in weathering of surface rocks, in the circulation of the ocean, and in the configuration of the continents, all of which in various combinations set the conditions that determined the climates of the past.

5. There are important advantages to fossil fuels as energy sources as compared to wood. Most important is their energy density. To supply the energy contained in a railroad car loaded with wood as opposed to coal, we would need to commandeer at least four or five additional rail cars. Furthermore, there are important advantages in having access to energy not only in solid but also in liquid and gaseous form. You couldn't imagine, for example, driving your car with either wood or coal (although one of the earliest automobiles, the Stanley Steamer, was in fact fueled by wood). Energy in gaseous or liquid form can be transported by pipeline over significant distances at relatively modest energy cost (as least in comparison with the cost of moving either wood or coal).

4

Human-Induced Climate Change

WHY YOU SHOULD TAKE IT SERIOUSLY

AS THE TITLE suggests, the objective of this chapter is to outline the reasons why you should take the issue of human-induced climate change seriously. There is no doubt that the concentrations of a number of the key components of the atmosphere, so-called greenhouse gases with the potential to alter climate, are higher now than they have been at any time over at least the past 800,000 years, probably much longer. Furthermore, there is no question but that diverse forms of human activity are largely responsible for the recent increase in the concentrations of these gases. The increase began several hundred years ago and, unchecked, is likely to continue into the indefinite future. This alone should justify a cautionary response with respect to the prospects for human-induced climate change. But, as we shall see, the issue is even more complicated.

The warming impact of the rising concentration of greenhouse gases has been offset (muted) until recently by cooling resulting from small particles formed in the atmosphere as byproducts of conventional forms of air pollution. These particles, referred to as aerosols, have an important negative impact on public health: when respired, they persist in your lungs and can enter the bloodstream, triggering a variety of serious cardiovascular and respiratory problems. They contribute also to the phenomenon of acid rain responsible for killing freshwater fish and damaging plant life more generally. Societies have taken steps with some success to reduce emissions of these offending chemicals. The downside is that we are now beginning to experience acceleration in the pace of climate change, a circumstance for which we are ill prepared.

I begin with a discussion of the energetics of the global climate system, the situation that would apply if energy absorbed from the sun on a global scale were balanced precisely by energy returned to space in the form of long-wavelength (infrared) radiation, following with an account of what we understand when we talk about the greenhouse effect, the role played by specific components of the atmosphere in limiting transmission of infrared radiation to space. As I shall discuss, these gases serve as insulation for the Earth (they shield the Earth's surface from the deep freeze of outer space). They are responsible in the present climate system for an increase in the average temperature of the planet's surface by about 40°C. In the absence of greenhouse gases, the Earth would be frozen over from equator to pole. None of this is particularly controversial.

I continue with a presentation of the evidence that the concentration of the key greenhouse gases is increasing, including an account of the factors responsible for this increase and a summary of evidence that the planet is warmer now than it has been for at least the past 140 years. Global surface temperatures have increased. The ocean is gaining heat and the planet at this point is clearly out of equilibrium: it is absorbing more energy from the sun than it is returning to space. I follow with an account of what we might expect from this out-of-equilibrium planet, including a summary of recent anomalous weather events—floods, droughts, heat waves, fires, and violent storms—an ominous preview for what may lie ahead. The chapter concludes with a summary of the key points. Given the critical importance of the climate issue for the central theme of the book—that human-induced change is real and we need to do something about it—I elected to include an extensive series of more detailed technical notes at the end of this chapter.

ENERGETICS OF THE GLOBAL CLIMATE SYSTEM

A representation of the energy budget of an idealized equilibrium global climate system is presented in Figure 4.1. The left-hand portion of the figure summarizes the fate of the energy that reaches the Earth from the sun, mainly in the visible.[1] The right-hand side illustrates the fate of long-wave, infrared, radiation, the primary mechanism by which heat absorbed by the planet is returned to space. On average, the Earth is exposed to 342 watts of solar energy for every square meter of its surface area, 342 W m^{-2} (as discussed in Chapter 2, a watt is a measure of power, energy per unit time). According to the data in Figure 4.1, 31.3% of the energy reaching the Earth from the sun is reflected back to space, from a combination of clouds and particulate matter (aerosols) suspended in the atmosphere (22.5%) and from the surface (8.8%). The fraction of incident (visible) solar energy returned to space is referred to as the Earth albedo: bright surfaces are reflective (high albedo), and dark surfaces are absorptive (low albedo).[2]

Some 49.1% of the solar energy intercepted by the Earth is absorbed at the surface, accounting for 71.5% of the total energy captured by the planet per unit time (168 W m^{-2}). A portion of the energy absorbed at the surface (14.3%) is transferred to the atmosphere by what are indicated in the figure as thermals, basically through contact of cold air with a warmer surface. A somewhat larger fraction (48.4%) is used to evaporate water, mainly from the oceans (accounting for as much as 33% of the total energy absorbed by

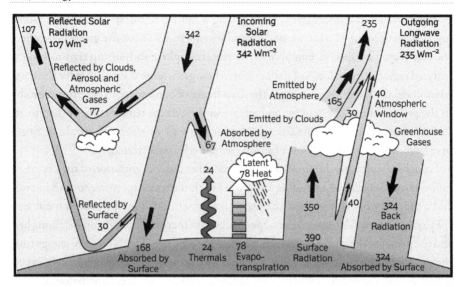

FIGURE 4.1 Estimate of the Earth's annual and global mean energy balance. Data quoted in units of watts per square meter, W m⁻².
(*Source:* IPCC 2007)

the Earth). Evaporation of water is balanced essentially in real time by precipitation (the atmosphere has limited capacity to store water vapor). What this means is that if we know the global rate at which energy absorbed at the surface is used to evaporate water, we can calculate the global average rate for precipitation. The data for evaporation in Figure 4.1 would imply a globally averaged rate for precipitation (expressed as liquid, as rain) of 117 cm or 46 inches per year.[3]

As indicated in the figure, energy is communicated to the atmosphere at an average rate of 235 W m⁻², of which 67 W m⁻² represents the source contributed directly by sunlight, with the balance, 168 W m⁻², associated directly and indirectly with transfer from the surface (the combination, as indicated in the figure, due to thermals, evaporation, and net exchange of infrared radiation with the surface).[4] The picture of the energy cycle presented in Figure 4.1 assumes an exact balance between energy sources and sinks for the globe: net input of 235 W m⁻² from the sun is offset precisely by emission of 235 W m⁻² in the form of infrared radiation to space. This is clearly an idealization. It provides nonetheless a useful representation of the key mechanisms for transfer and disposal of energy in a hypothetical equilibrium climate system.

The bulk of the infrared radiation emitted to space originates from a level of the atmosphere where the temperature is about −18°C, from an altitude of about 5 km. If the atmosphere were composed exclusively of O_2 and N_2 (these gases account for 99% of the present atmosphere), the atmosphere would be transparent to infrared radiation and radiation emitted by the surface would be transferred directly to space. The Earth would be frozen over in this case—from equator to pole.[5] What makes the difference is that the atmosphere contains trace quantities of gases that are able to absorb infrared (heat) radiation, notably H_2O (water vapor), CO_2 (carbon dioxide), CH_4 (methane), O_3 (ozone), N_2O (nitrous oxide), and a host of other minor absorptive species referred to collectively as greenhouse gases. These gases serve as traps for heat and are primarily responsible for the relatively mild temperatures that define average conditions at the Earth's surface today. The most important of the greenhouse gases in the present atmosphere is water vapor. The abundance of water vapor depends, however, on the state of the climate system: if the climate is cold, the capacity of the atmosphere to hold H_2O is severely limited; conversely, if the climate is warm, the abundance of atmospheric H_2O would be relatively high and the role of the greenhouse would be amplified accordingly. In this sense, H_2O may be considered as a derivative or secondary greenhouse agent. Its abundance, and thus its contribution to the overall greenhouse effect, depends ultimately on the heat-trapping capacity of the other greenhouse gases, notably CO_2.

The level in the atmosphere from which infrared radiation escapes to space depends on the abundance of overlying absorbing gases (or clouds or aerosols). What happens if we were to increase instantaneously the abundance of these absorbing species? In this case, radiation would be unable to make it directly to space from the original escape region: it would be reabsorbed. The escape level would have to move higher in the atmosphere as a consequence. But the temperature there would be lower than the temperature at the original emission level. What that implies is that emission of radiation to space would be reduced (emission depends on the fourth power of the emitting temperature). All other factors being equal (no changes in the net energy absorbed from the sun), the Earth would absorb more energy from the sun that it could emit to space. It would therefore have to warm. Conversely, an increase in the abundance of reflective aerosols (discussed later) would result in a decrease in energy absorbed from the sun. The planet would have to cool in this case. The key question concerns the ultimate change in global temperature and consequently climate resulting from a particular disturbance and the manner in which the climate system might be expected to adjust over time to a specific imposed change in the planetary energy budget.

Consider the effect of a sustained rise in the concentration of a greenhouse gas such as CO_2. As discussed earlier, the immediate impact would be to increase the (net) energy absorbed by the planet. This extra energy would be expected to prompt additional changes in the planetary energy budget, changes that could be either positive or negative (amplifying or attenuating the response to the initial disturbance). An increase in the abundance of water vapor would be projected, for example, to enhance the warming resulting from the increase in CO_2 (recall that H_2O is an important greenhouse gas). The feedback would be positive in this case. On the other hand, a higher abundance of H_2O could trigger an increase in global cloud cover. The decrease in net energy input resulting from this brighter, more reflective planet would serve to offset (compensate), partially at least, warming triggered by the initial rise in the abundance of water vapor. The feedback would be negative in this case. It is reasonable, though, to expect that the overall impact of an increase in the concentration of greenhouse gases on the planetary energy budget should be positive and that the temperature of the atmosphere would have to increase in order to restore energy balance (to increase emission of infrared radiation to space).

Climate scientists define the ultimate equilibrium response of globally averaged surface temperature to a specific imposed energy disturbance in terms of what they refer to as the climate sensitivity. Charney et al., in a classic early study commissioned by the US National Academy of Sciences in 1979, concluded on the basis of fundamental scientific principles and model results that a change in globally averaged energy input of 1 W m^{-2} (radiative forcing as defined later) would result in a change in globally averaged surface temperature of 0.75 ± 0.375°C. A more recent study by Hansen and Sato (2012), based on a consideration of climate data extending back to 800,000 years before present, narrowed the range of uncertainty, reporting a sensitivity of 0.75 ± 0.15°C. An instantaneous (sustained) doubling of the concentration of CO_2 with respect to the level present in the preindustrial environment (280 parts per million) would be predicted to trigger an imbalance (increase) in global energy input of about 4 W m^{-2}.[6] Adopting Hansen and Sato's estimate for climate sensitivity, this would result (eventually) in an increase in global average surface temperature by between 2.5°C and 3.5°C.

The atmosphere is expected to respond on a somewhat delayed basis to a change in the net input of energy to the planet. A portion of the energy excess resulting from an increase in the concentration of greenhouse gases will be absorbed by and used to warm continental landmasses. A larger fraction will be deployed to heat the ocean. Models suggest that approximately 40% of the anticipated equilibrium change in surface temperature may be realized in as little as 5 years with land areas warming up more rapidly than the ocean, reflecting the much greater heat capacity of the latter. Hundreds of years or even longer will be required for the system to evolve to its new equilibrium state, reached finally only when emission of infrared radiation to space has increased to the extent required to restore global energy balance (to compensate for the imbalance in energy triggered by the initial change in composition). The delay reflects the relatively sluggish rate at which heat can be transferred from upper to deeper regions of the ocean. The response to a decrease in energy input would be similarly delayed: a large volume of ocean water would have to cool in this case before the temperature of the planet's surface and atmosphere could adjust to the lower values required to restore equilibrium. What this means is that we are likely to have to live for a long time (decades at least) with the climatic consequences of the changes in the composition of the atmosphere for which we are responsible not only today but also for an extended time in the past. The greater the climate sensitivity, the longer the time required for the system to transition to the new equilibrium state.[7]

GREENHOUSE GAS RECORD

Without question, the concentration of greenhouse gases is higher now than it has been for at least the past 800,000 years. How do we know this and how can we be sure that humans are responsible for the rapid increase that has taken place over the past several hundred years? Measurements of gases trapped in polar ice provide an invaluable record of long-term changes in the composition of the atmosphere.

Results for CO_2, CH_4, and N_2O are displayed in Figure 4.2. The lower two curves in the figure present surrogate records for the corresponding changes in global climate.[8] Concentrations of CO_2 varied between about 180 parts per million (ppm) and 290 ppm over the entire record included in the figure. Concentrations of CH_4 ranged from about 400 parts per billion (ppb) to about 700 ppb with N_2O fluctuating between about 200 and 280 ppb (the record is less complete in the case of N_2O). Concentrations of all three gases were typically low during ice age conditions and high during periods of relative warmth. At no time did the concentrations of any of these gases approach the levels that apply today (approximately 400 ppm for CO_2, 1,850 ppb for CH_4, and about 320 ppb for N_2O).

The contrast between past and present is emphasized even more dramatically with the data presented in Figure 4.3, summarizing the changes that have taken place over the past 20,000 years since the end of the last ice age. The increases in concentrations over the past few hundred years are displayed as essentially vertical lines on the right-hand sides of the curves in Figure 4.3. Panels a, b, and c in the figure include estimates (the right-hand axes) for what the Intergovernmental Panel on Climate Change (IPCC)

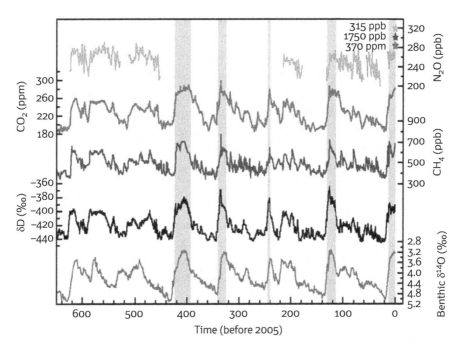

FIGURE 4.2 Changes in concentrations of the greenhouse gases CO_2, CH_4, and N_2O. (*Source:* IPCC 2007)

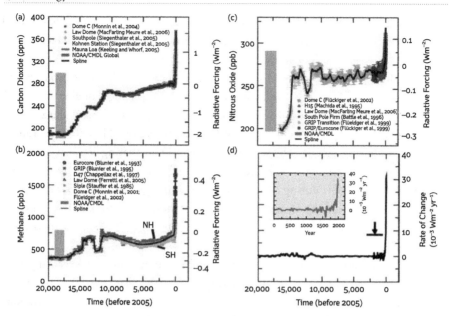

FIGURE 4.3 Concentrations and contribution to radiative forcing over the past 20,000 years for (a) CO_2, (b) CH_4, and (c) N_2O. Panel (d) illustrates the rate of change of radiative forcing. (*Source:* IPCC, 2007)

refers to as the radiative forcing associated with the observed compositional changes, a useful if approximate estimate of the significance for the change in net energy input to the planet occasioned by the observed changes in the concentrations of the greenhouse gases included in the figure.[9] The values presented in panel d, identified as the rate of change of radiative forcing, were computed on the basis of the time rate of change of the sum of the contributions to forcing indicated in panels a–c and are intended to indicate the significance of the change in radiative forcing experienced per year as a function of time. Hansen et al. (2011) estimate a value of approximately 3 W m^{-2} for the effective net radiative forcing that had developed by 2003 in response to the increase in the concentration of greenhouse gases observed over the past several hundred years—since the dawn of the modern industrial age.[10]

Estimates by Hansen et al. (2011) for the year-to-year changes in radiative forcing associated with the changing concentrations of greenhouse gases over the past 50 years are displayed in Figure 4.4. Notable is the rapid increase in annual incremental forcing inferred to have taken place between 1960 and 1990, followed by the decrease after 1992, the partial recovery from 1992 to 1998, and the relatively constant level that has applied over the past decade. The decrease in the late 1980s and early 1990s was due to a temporary decrease in the rate of growth of CO_2 combined with a slowdown in the growth of CH_4 and the chlorofluorocarbons (CFCs), the latter reflecting the success of the Montreal Protocol implemented in 1987 to protect the stratospheric ozone layer. Present annual incremental forcing averages about 0.04 W m^{-2}. Forcing at this level if

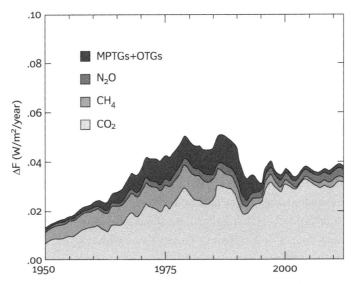

FIGURE 4.4 Year-to-year changes in radiative forcing associated with greenhouse gases over the past 50 years.
(*Source:* Hansen et al. 2013)

sustained for a decade (as appears to be the case for the most recent decade) would be sufficient to cause an eventual increase in global average surface temperature by about 0.3°C (assuming the value for climate sensitivity noted earlier, 0.75°C per W m^{-2}). Much of this increase, however, will take place in the future, reflecting the delay imposed by the thermal inertias of the ocean and to a lesser extent the land surface. Incremental forcing over the earlier period (as much as 3 W m^{-2}) is more than sufficient, however, as we shall see, to account for the observed increase in globally averaged surface temperature and for the additional heat observed to have accumulated in the ocean.

Can there be any doubt that human activities are responsible for the large changes in the concentrations of greenhouse gases observed over the recent past? What are the chances that the changes observed over the past few hundred years might be interpreted as a natural excursion, a once-in-an-800,000-year anomaly? The answer, of course, is not likely. And, in fact, we actually understand, at least in general terms, the factors responsible for these recent changes. They relate in part to our increased reliance on fossil fuels as a primary energy source, in part to stresses imposed by demand for the resources and technology required to feed a population that has now for the first time in history surpassed the 7 billion mark.

The increase in the abundance of CO_2 is due primarily to emissions associated with combustion of coal, oil, and natural gas. This source accounted for the addition of 89.3 billion tons of CO_2 to the atmosphere between 1995 and 2007 (close to 13 tons for every man, woman, and child on the planet), of which 29% was transferred to the ocean, 15% was absorbed by the biosphere, and 57% was retained by the atmosphere (Huang and

McElroy 2012). Absent a major shift in global energy policy, we can anticipate that the current trend in accelerating emissions is likely to continue for the foreseeable future. The prospect is for the abundance of CO_2 to increase by as much as a factor of 2 over the next several decades as compared with the level that applied in the preindustrial period (to rise to values in excess of 560 ppm).

Methane is produced by microbes feeding on organic material under anaerobic (low-oxygen) conditions. Organic-rich swamps are responsible for a significant natural source (in fact, methane is sometimes referred to as swamp gas). Rice paddy fields mimic conditions in natural swamps and are implicated in further production. Domesticated animals provide an additional source.[11] The fossil fuel sector is also involved. The source in this case includes releases in conjunction with the extraction and combustion of the fossil fuels (notably coal and natural gas), emissions relating to incomplete flaring of gas by the oil industry, and inadvertent losses associated with leaks in the gas distribution system. Large quantities of organic carbon are stored at present in frozen soils at high latitude. There is concern that future warming could melt the ice in these soils, turning them into potential major new sources of CH_4 (and CO_2), accelerating the conditions for further warming. Given the complexity of potential sources, including the possibility of important climate-related feedback, in addition to uncertainties in the future efficiency with which the gas is removed from the atmosphere,[12] it is difficult to project the future course of atmospheric CH_4.

It is similarly challenging to forecast the future of N_2O. Without question, microbial processing of nitrogen represents the dominant source for this greenhouse agent. The rise in the concentration of N_2O over the recent past is associated most likely with major additions of nitrogen in the form of chemical fertilizer, together with the processing of increased flows of organic nitrogen contributed by a combination of nitrogenous wastes. These wastes are produced not only by the increasing population of people inhabiting our planet but also by the growing population of chickens, cows, sheep, and other animals we nurture to satisfy our ever-rising demand for high-grade protein and related dairy products. Introduction of fresh sources of what we refer to as fixed nitrogen in the form of fertilizer applied to fields to promote enhanced production of the food crops accounts for an additional important source.[13] It is doubtful, given the complexity of the biologically mediated global nitrogen cycle, that we can seriously curtail the factors responsible for the present and projected future rise in N_2O. It is for this reason that I have chosen in this volume to focus on the sources of greenhouse gases that relate most directly to our use of fossil fuels, and that can potentially be addressed by inventive policy, specifically CO_2 and CH_4.

INCREASES IN GLOBAL SURFACE TEMPERATURES

Figure 4.5 presents a summary of the changes in global average surface temperature observed over the past 130 years—from 1880 to 2010. The data are displayed here in terms

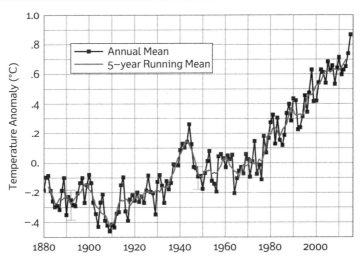

FIGURE 4.5 A summary of the changes in global average surface temperature observed over the past 130 years—from 1880 to 2013. Uncertainties are indicated by light colored bars. (http://data.giss.nasa.gov/gistemp/graphs_v3/)

of the changes that have occurred since 1880, referenced with respect to the average of temperatures observed between 1951 and 1980 (the zero reference level in the figure). Temperatures were relatively constant from 1880 to about 1920. They increased by about 0.5°C between 1890 and 1945, flattened out or even declined between 1945 and 1975, and have risen since by about 0.6°C. There is no question that the Earth, at least at its surface, is warmer now than at any time over the past 130 years. The rise in temperature over the entire record is about 1.0°C with approximately 70% of the increase having taken place over the past 40 years.

ROLE OF AEROSOLS

The increase in globally averaged surface temperatures observed since 1880 is actually less than would have been anticipated given the contribution expected from the observed increase in the concentration of greenhouse gases. As indicated earlier, particulate matter in the atmosphere (aerosols) can provide a potential offset to greenhouse-induced warming. Reflection of sunlight by light-colored aerosols can result in enhanced reflection of sunlight (increase the planetary albedo), contributing to a decrease in energy absorbed from the sun: this is what IPCC (2007) refers to as the direct effect of aerosols. Aerosols can serve also as nuclei for condensation of water vapor, prompting in this case an increase in the number of cloud particles and consequently cloud reflectivity, leading to a further reduction in the input of energy to the planet.[14] IPCC (2007) refers to this as the indirect effect of aerosols. Natural sources of aerosols include salt particles formed by evaporation of sea spray, particles formed by photochemical processes in the atmosphere, and a variety of different forms of wind-blown dust, including but not limited to light-colored particles raised by windstorms in desert regions (a source of negative radiative forcing). Human-related (anthropogenic) sources encompass a variety of forms of particulate matter produced as a result of emissions of sulfur dioxide (SO_2), nitrogen oxides (NO_x), and ammonia (NH_3), together with a range of different forms of organic carbon, including what is referred to as black carbon or soot. The particulate matter formed as a consequence of the emissions of SO_2, NO_x, and NH_3 is light in color and generally effective in promoting condensation when conditions are appropriate. Thus, these emissions are expected to contribute a source of negative radiative forcing. Forcing associated with black carbon, as the name suggests, is positive. The weight of the evidence suggests, however, that, at least on a global scale, the cooling impact of the light-colored, condensation-favoring, emissions of SO_2, NO_x, and NH_3 is greater than the potential for warming contributed by black carbon.[15]

In contrast to the relative confidence attached to estimates for the positive radiative forcing associated with greenhouse gases, estimates for the negative forcing associated with anthropogenic aerosols are regrettably uncertain. IPPC (2007), drawing on a variety of literature studies, recommended a value of -1.2 W m^{-2} for the cumulative forcing by aerosols since 1750, with error bars reported to include values as great as -2.7 W m^{-2} or as small as -0.4 W m^{-2}. Underlining the significance of the wide range of uncertainty in the aerosol contribution quoted by the IPCC, it should be noted that a value as great as -2.7 W m^{-2} would be sufficient to offset the bulk of the positive forcing attributed to the increase in greenhouse gases. Murphy et al. (2009) used an empirical approach to constrain the permissible range of values for this important parameter. Using space-based measurements of the planetary energy balance from the Earth Radiation Budget Experiment System (ERBE) and the Clouds and the Earth's Radiant Energy System (CERES) missions combined with ocean heat content data (discussed later), they concluded that aerosol forcing was responsible for negative forcing of -1.8 W m^{-2} in 2000. An independent analysis by Hansen et al. (2011) reported a value of -1.6 ± 0.3 W m^{-2} for 2010 with an average of -1.1 ± 0.3 W m^{-2} for the period 1970 to 2000. There are reasons to believe, however, as discussed later, that the moderating influence of aerosols on climatic warming may have waned somewhat in recent years and that the capacity of aerosols to offset warming from greenhouse gases may be more limited in the future.

Emissions of SO_2, displayed in Figure 4.6, offer a useful surrogate for the trend over time in the global source of (anthropogenic) aerosols. A breakdown of individual contributions to these emissions is displayed in Figure 4.7. The rapid increase in global emissions from about 1950 to 1970 offers a potential explanation for the pause observed over this time interval in the otherwise inexorable rise in global average surface temperatures. Global emissions peaked in the early 1970s, reflecting primarily policy measures implemented in the United States and Europe to minimize the environmental damage associated with acid rain (see Fig. 4.7 for the temporal trend in US emissions). The increase in global average surface temperatures resumed subsequently. Emissions from China and from shipping (Fig. 4.7) now constitute the most important, most rapidly growing,

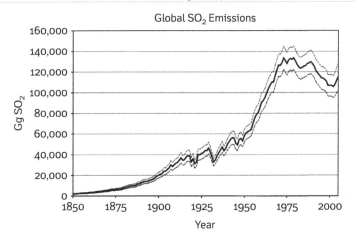

FIGURE 4.6 Global sulfur dioxide emissions from fuel combustion and process emissions with central value (solid line) and upper and lower uncertainty bounds (dotted lines).
(*Source:* S. J. Smith et al. 2011)

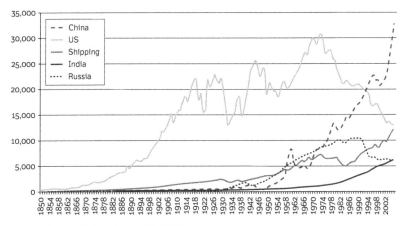

FIGURE 4.7 Top five SO_2 emitters (Gg SO_2).
(*Source:* Smith et al. 2011)

sources of sulfur emissions, responsible in combination for the modest increase in the global source indicated since 2000 in Figure 4.6. There are indications, however, that Chinese emissions may have decreased more recently as a result of measures taken by the Chinese government to address problems associated with local and regional (conventional) air pollution. Furthermore, as discussed by Lu et al. (2012), emissions from the United States have decreased precipitously over the past 4 years, reflecting the switch from coal to natural gas in the power sector prompted by the recent fall in prices for natural gas, which has resulted in selective idling of the most inefficient (oldest) coal-fired power plants that were generally not equipped to remove sulfur. As noted earlier, the ongoing downward trend in emissions from China and the United States raises the possibility for a significant increase in net radiative forcing in the immediate future: in fact, we may already be experiencing the impact of this increase, as discussed later.

NATURAL SOURCES OF CLIMATE VARIABILITY

In addition to the longer term impact of human-induced changes in greenhouse gases and aerosols, the climate system may be expected to respond to a range of influences that may be considered natural in origin, variability reflecting the essential non-linearity of the underlying physics and the interaction of the different parts of the climate system—atmosphere, ocean, ice, and ecosystems. Changes may arise on a variety of time scales ranging from days to months in the case of weather, to years or even hundreds of years in response to changes in the ocean. Examples of some of the more persistent changes affecting climatic conditions in the tropics and over large regions of the northern hemisphere include the El Nino–Southern Oscillation (ENSO), the Madden Julian Oscillation (MJO), the Atlantic Multidecadal Oscillation (AMO), the North Atlantic Oscillation (NAO), the Arctic Oscillation (AO), and the Pacific Decadal Oscillation (PDO) with similar quasi-regular fluctuations observed in the southern hemisphere. An interesting, though unresolved, question concerns the potential significance of the interplay between natural variability as exemplified by these various fluctuations and human-induced change in regulating at least the short-term expression of the global climate system.[16]

The El Nino–Southern Oscillation (ENSO) phenomenon offers a relatively well-understood example of multiyear variability that arises as a consequence of two-way interactions between the atmosphere and ocean in the tropical Pacific. The extremes of ENSO are referred to as El Nino and La Nina. El Nino (the warm phase) is identified with unusually warm waters in the surface tropical Pacific extending from the dateline to the coast of South America. Upwelling of nutrient-rich water from lower colder regions of the ocean off the coast of South America and flow of this water across the tropical ocean define the very different conditions that prevail during La Nina (the cold phase). Global average surface temperatures are slightly elevated during El Nino and depressed during La Nina. The impact of ENSO is experienced over extensive regions of the globe, most particularly in the tropics and subtropics. El Nino is associated with drought over Indonesia, Australia, and southern Africa, and with unusually heavy rains over Peru and Ecuador. Meteorological conditions during La Nina reflect essentially a mirror image of those observed during El Nino (heavy rains over Indonesia, Australia, and South Africa; drought over Peru and Ecuador). The La Nina condition, which has prevailed over the past several years, as indicated in Figure 4.8, may have been responsible, at least in part, for the devastating drought and out-of-control fires experienced most recently over large regions of the American Southwest.[17]

Muller et al. (2012) concluded that the decadal variability of global land surface temperatures over the past 60 years is correlated significantly (correlation coefficient of 0.65)

Human-Induced Climate Change: Why You Should Take It Seriously   47

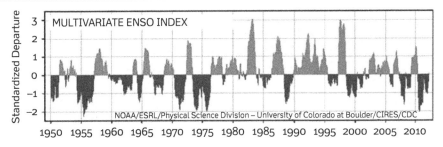

FIGURE 4.8 ENSO Index.
(*Source:* NOAA/ESRL)

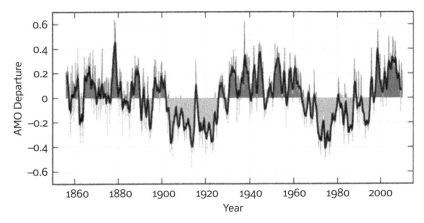

FIGURE 4.9 Monthly values for the AMO Index, 1856–2009.

with the variability of the AMO, the multidecadal Atlantic Oscillation.[18] The AMO is associated with a more or less coherent pattern of variability of sea-surface temperatures in the North Atlantic. Monthly values for the departure of the AMO from the mean are displayed for the interval 1856–2009 in Figure 4.9. The index varies on a time scale of between 60 and 90 years and has been associated with changes in the meridional (latitudinal) overturning of the Atlantic.[19] Booth et al. (2012) have argued that changing patterns of sulfur emissions may also have played a role at least over the past century. Knudsen et al. (2011) present evidence that the AMO has been a persistent feature of north Atlantic climate variability for at least the past 8,000 years. They point to a series of studies linking the AMO to changes in precipitation over North America, droughts in the Sahel, variability in northeastern Brazil rainfall, the frequency and intensity of tropical hurricanes, and potentially even to changes in the interhemispheric transport of heat.

A comparison of trends in temperatures over land as compared to over the ocean is displayed for the past 60 years in Figure 4.10. Notable here is the increase in land versus ocean surface temperatures since about 1990. This is precisely what one would expect if, as suggested, there has been a recent decrease in greenhouse offsetting aerosols: the

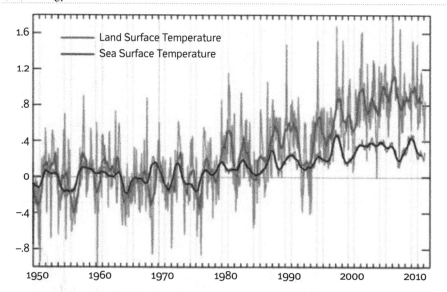

FIGURE 4.10 Monthly (thin lines) and 12-month running mean (thick lines) global land and sea surface temperature.
(*Source:* NASA/GISS)

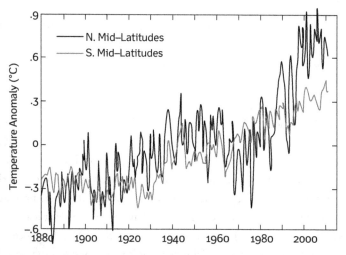

FIGURE 4.11 Temperature change for mid-latitude bands (12-month running mean).
(*Source:* NASA/GISS)

additional warming should be experienced first on land and later in the ocean, reflecting the greater heat capacity of the latter. We would expect the additional warming to show up also more immediately in the northern hemisphere as compared to the southern since the land area of the former exceeds that of the latter and since the radiative impact of aerosols should be much greater in the north (where most of the industrial sources are concentrated). Consistent with this expectation, the increase in temperatures observed

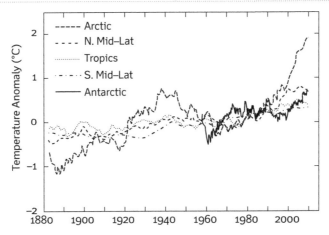

FIGURE 4.12 60-month running mean temperature changes in five zones as indicated. (*Source:* NASA/GISS)

at northern mid-latitudes over the most recent period has been generally greater than reported at southern mid-latitudes (Fig. 4.11), greatest of all in the Arctic (Fig. 4.12).

INCREASES IN OCEAN HEAT CONTENT

If the trends in surface temperatures displayed earlier can be interpreted as persuasive evidence for a change in global climate, data from the ocean provide even more convincing proof that the energy budget of the Earth is no longer in balance. *Without question, the planet is now gaining energy: energy absorbed from the sun indisputably exceeds that returned to space in the form of infrared radiation.* The increase in the heat content of the ocean clearly attests to this fact. It has been going on for at least the past 50 years, arguably much longer, probably for more than a century.

There has been an impressive increase in measurements of the heat content of the ocean (an indication of total stored energy) over the past several decades attributed largely to the success of the international ARGO float program. The ARGO program involves a series of drifting robotic probes, deployed essentially worldwide, designed to record measurements of ocean temperature and salinity to depths as great as 2,000, more realistically to about 1750 m. The probes surface every 10 days, relaying their data by satellite to receiving stations around the world, with the information made available without restriction subsequently to the global community of ocean scientists. Some 3,500 of these probes were deployed and were operational as of March 2012.

A summary of data on ocean heat content reported by Trenbreth and Fasullo (2012), essentially an update of results presented by Levitus et al. (2012), is displayed in Figure 4.13. A notable feature of this presentation is the recent increase in the heat content of the ocean at deep levels (below 700 m) as compared to the more muted change observed at shallower levels (above 700 m). The rate of increase in ocean heat content to the

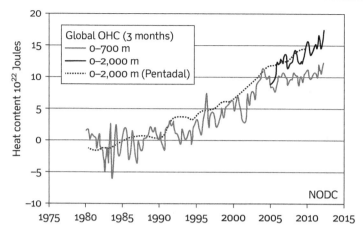

FIGURE 4.13 The global ocean heat content in 10^{22} J from NODC (NESDIS, NOAA), updated from Levitus et al. (2012). The different curves show three monthly average values for 0–700 m and 0–2,000 m. The dashed curve is the pentadal (running 5-year) analysis for 0–2,000 m for which, in the 1980s, the 2 sigma standard error is about $\pm 2 \times 10^{22}$ J, decreasing to $\pm 1 \times 10^{22}$ J in the early 1990s, but increasing in the late 1990s until it decreases substantially to about $\pm 0.5 \times 10^{22}$ J in the Argo era. The reference period is 1955–2006.

lowest levels sampled by ARGO averaged about 0.6 W m^{-2} between 1993 and 2011 (expressed with respect to total planetary surface area).[20] Hansen et al. (2011) adopted a value of 0.51 W m^{-2} (again globally averaged) for the interval 2005 to 2010 with an uncertainty of 0.12 W m^{-2} attributed to differences in treatments of the ocean data. They associated a somewhat higher value (0.625 W m^{-2} with a similar uncertainty range) with the more extended period 1993–2008, consistent with the modest decrease in the growth rate of the 0–700 m ocean heat content observed most recently (Levitus et al. 2012). The decrease in solar luminosity during the most recent solar minimum may have contributed, as suggested by Hansen et al. (2011), to the slowdown in heat uptake by upper levels of the ocean (above 700 m) since 2004, as indicated by the data in Figure 4.13.[21]

Recent reanalysis of data from the British Challenger expedition (1872–1876) (Roemmich et al., 2012) suggests that the heat content of the ocean may have increased by as much as a factor of 2 over the past 135 years, with approximately half of the increase taking place prior to the 1950s.[22] The obvious conclusion: the Earth has been out of energy balance for much of the past century, consistent with the trend in global average surface temperatures indicated in Figure 4.5.

WHAT WE CAN EXPECT FROM PLANETARY WARMING

In brief, we can anticipate an increase in the incidence of floods and droughts, storms that become increasingly more destructive, weather systems for which our infrastructure is ill prepared, an increase in the incidence of extreme heat waves, a decrease in the extent and thickness of sea ice in the Arctic, deserts migrating to higher latitude, and all of this combined with a damaging rise in global sea level.

As indicated earlier, precipitation on a global scale is controlled ultimately by evaporation, mainly from the ocean. The atmosphere is warmer now than it has been in the past. Since the capacity of the atmosphere to hold water vapor is linked fundamentally to its temperature, there can be little doubt that the abundance of atmospheric water vapor has increased and that it is likely to rise further in the future with additional warming. Assume for the moment that there is no significant change in global evaporation (and thus precipitation). When conditions favor precipitation (rising motion), it is likely that the intensity of the resulting precipitation will increase. If global precipitation is conserved (or increased only modestly), this means that it must rain or snow less elsewhere. What we may expect then is weather that is less predictable and generally more extreme, increased incidences of floods in some regions compensated by droughts in others.[23] The energy of a storm is derived in no small measure from the change of phase of water (conversion of vapor to either liquid or ice). With an enhanced supply of water vapor, we can expect not only extremes in the global hydrological cycle in the future but also for storms to become more powerful, feeding on this additional source of energy. Recent experience is consistent with this expectation.

Loss of life and serious damage to property are associated generally with episodes of high wind, intense precipitation, or both. Of particular interest is the implication of planetary warming for violent storms—for hurricanes, major cyclones, and tornadoes, for example. Emanuel (2007) has argued that, while there may be no reasons to expect an increase in the number of hurricanes (or typhoons as they are referred to in the Pacific), it is likely that these storms, feeding off higher ocean surface temperatures, will become more energetic in the future. A similar conclusion holds for major cyclones. On the other hand, despite public perception to the contrary, there is no evidence that tornadoes have become more frequent or more damaging recently, nor is there evidence that they have expanded their range. Scientific evidence is simply lacking on this score.

The United States experienced its warmest year on record between June 2011 and June 2012. Each of the consecutive months ranked among the warmest third of months in the historical record. The odds that this should reflect a random natural fluctuation are rated at less than one in a million (http://www.wunderground.com/blog/JeffMasters/comment.html?entrynum=2149). The global pattern is similar. Maximum temperatures have been rising: the good news is that nighttime and winter temperatures are not as cold as they used to be. More people die, however, as a result of exposure to high temperatures as compared to exposure to conditions that are unusually frigid (in the latter case, you can always protect yourself by dressing warmly). The National Weather Service has developed an index taking into account the effects of temperature and humidity on human comfort (or discomfort). Prolonged exposure to temperatures in excess of 100°F (Category II) with moderate exercise, even at relatively low levels of humidity, may be expected to result in serious heat stroke. Exposure under similar circumstances to temperatures above 120°F (Category I) could prove fatal. Plants and animals are similarly vulnerable.

There has been a notable decrease in the coverage of summer ice in the Arctic Ocean in recent years, as indicated in Figure 4.14. The decline in areal extent of sea ice in this region in summer has been accompanied by a decrease in the presence of multiyear ice (the ice is getting thinner). Ice coverage maximizes typically in mid to late March, with a minimum in September. Ice extent in August of 2007, more than a month before the end of the traditional summer melt season, set a record low, paving the way for the first time for shipping from Asia to Northern Europe to proceed unimpeded through the Northwest Passage.[24] By 2011 the ice extent in September had declined to 4.6 million km² as compared to an average of 7.0 million km² observed between 1979 and 2000 (http://earthobservatory.nasa.gov/Features/WorldOfChange/sea_ice.php/). A series of positive feedbacks conspire to amplify the impact of warming in the Arctic. Reduction in the extent of sea ice coverage permits relatively warm ocean water to communicate directly with the atmosphere. In addition, the decrease in reflective ice cover allows for enhanced absorption of sunlight by the ocean, particularly in summer. The result is a self-reinforcing increase in warming, with implications not only for the Arctic Ocean but also for climate in the polar region more generally and potentially also for weather systems at lower latitudes.

There is evidence for a recent expansion of what is known as the Hadley circulation, the climatic system that dominates meteorological conditions in the tropics and

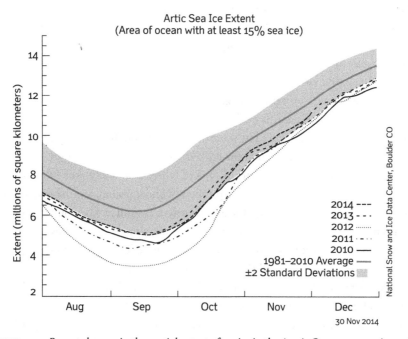

FIGURE 4.14 Recent changes in the spatial extent of sea ice in the Arctic Ocean: comparison of data from 2010 to 2014 with the average for 1981–2010. The 1981 to 2010 average is in dark gray. The gray area around the average line shows the two standard deviation range of the data.
(*Source:* http://nsidc.org/arcticseaicenews/)

subtropics. Moisture-laden air rises in the equatorial region. After losing most of its moisture due to precipitation, it moves to higher latitudes in both hemispheres. It sinks to the surface in the subtropics at latitudes between about 25° and 35° before returning to the equatorial source region through the lower branches of the circulation loops (defined by the trade winds). When the air sinks to the surface in the subtropics, it has already lost most of its moisture. Its temperature rises as its sinks to the surface, by as much as 9°C per kilometer of descent. The major desert regions of the world are generally co-located with the descending loops of the Hadley system. A poleward shift of the Hadley circulation system could cause these desert regions to transition to higher latitude: think of the Sahara extending across the Mediterranean into southern Europe or the southwestern desert of the United States moving north into the grain-producing region of the country. The observational data indicate that the Hadley system expanded by as much as 2 degrees of latitude between 1979 and 2005 (Fu et al. 2006; Seidel and Randel 2007). It is interesting to note that the observed expansion is greater than predicted by climate models, by as much as a factor of 10, raising the possibility that the models may actually underestimate the significance of human-induced climate change.

Changes in global sea level have been measured with significant precision over the past 19 years using instrumentation (altimeters) mounted on the US/French Jason-1, Jason-2, and Topex/Poseidon spacecraft. A summary of results from these missions is displayed in Figure 4.15 (http://sealevel.colorado.edu/). The increase in sea level over the period

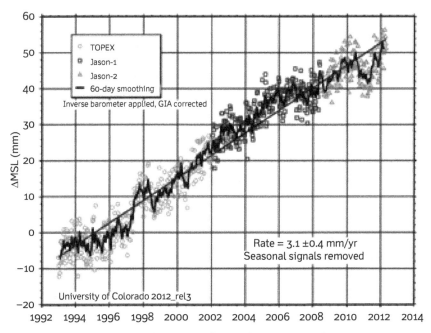

FIGURE 4.15 Global mean sea level time series (seasonal signals removed).
(*Source:* http://sealevel.colorado.edu/)

covered by the satellite observations has averaged 3.1 ± 0.4 mm per year, approximately twice the rate projected by IPCC. The increase in the temperature of the ocean (the resulting expansion of ocean waters) can account for about 30% of the observed change. A further 30% or so is attributed to retreat of worldwide mountain glaciers with the balance assigned to changes in the major ice sheets covering Greenland and Antarctica.

Measurements of the strength of gravity by the GRACE satellite (Velicogna, 2009) provide definitive evidence for the importance of the contributions from Greenland and Antarctica: loss of ice mass from Greenland increased from 137 billion tons per year in 2002–2003 to 286 billion tons per year in 2007–2009; loss from Antarctica grew from 104 billion tons per year in 2002–2006 to 246 billion tons per year in 2006–2009. In combination, Greenland and Antarctica were responsible for a rise in global sea level averaging 1.1 mm per year between April 2002 and February 2009. The fact that the rate of rise contributed by these sources appears to be increasing in time is cause for concern.

A major change in release of ice from these environments in the future could have the potential to alter sea level, not by centimeters but potentially by meters or even more. As reported by NASA, close to 97% of the surface ice covering Greenland melted over a 4-day period beginning on July 8, 2012, an event without recorded precedent (http://news.yahoo.com/nasa-strange-sudden-massive-melt-greenland-193426302.html). While melting of surficial ice is not expected to have a direct impact on the mass balance of ice on Greenland, it could have an indirect effect. Percolation of liquid into cracks in the ice and subsequent freezing (and expansion) could increase the rate at which ice is normally mobilized and transferred to the ocean.[25] A further possibility is that, if subjected to multiple episodes of such surface melting, melt water could eventually reach the base of the ice sheet, whence it might be able to make its way directly to the ocean.

UNUSUAL PATTERNS IN RECENT WEATHER

Thomas Friedman chose as title for his op-ed column published by *The New York Times* on September 13, 2011, the provocative question: "Is it weird enough for you?" In the article, he defined "global weirding" as "the hots get hotter; the wets get wetter and the dries get drier." The weather has certainly been weird over the past several years.

Pakistan suffered from back-to-back major floods in 2010 and 2011. More than 20 million people were displaced with damage estimated to exceed $50 billion. The State of Queensland in Northeastern Australia was visited by record rains in late December of 2010 and January of 2011. More than 25,000 homes and 5,000 businesses were inundated in Brisbane when two major rivers burst their banks. Russia experienced an unprecedented heat wave and drought in the summer of 2010. Temperatures in Moscow broke the 100°F mark on July 29, setting an all-time record. The heat and drought resulted in massive crop failures with losses from the drought and associated wildfires estimated to exceed $15 billion. The Horn of Africa suffered over the past few years through the worst drought in 60 years, prompting a major humanitarian disaster. And, in the fall of 2011, large parts of Thailand, including the capital Bangkok, were under water.

The United States has not been immune to these weather-related disasters. The National Climate Data Center of the US Department of Commerce provides an instructive summary of what they classify as weather/climate-related disasters over the past 31 years (damages from individual events in excess of $1 billion, 2001-adjusted dollars: total losses in excess of $750 billion). Losses for the first 9 months of 2011 were estimated at more than $45 billion. Persistent rainfall in the Ohio Valley (300% of normal) combined with melting snow pack resulted in historic flooding along the Mississippi River and its tributaries during 2011 spring and summer—damage in excess of $2 billion. Drought, heat waves, and wildfires in the southern plains/southwest dominated national news reports in the summer of 2011. Damage to range lands in this case required large numbers of cattle to be slaughtered as fodder was depleted and in some cases totally eliminated. Fire-fighting and fire-suppression costs exceeded $1 million per day: total losses of more than $9 billion. More than 340 strong tornadoes struck the central and southern states between April 25 and 30: 321 people died, 240 in Alabama alone. Costs for damage to property exceeded $9 billion. A further outbreak of tornados between May 22 and 27 led to 177 additional casualties. The city of Joplin, Missouri, was leveled by what was considered the most deadly tornado ever to strike the United States since modern tornado records began in 1950: losses from this outbreak of tornados were estimated to exceed $8 billion. And this is only a partial list of 2011 weird weather.

The situation had not much improved in 2012. A number of major agricultural countries, notably the United States, Canada, Mexico, Brazil, Argentina, Spain, Italy, Ukraine, Russia, and India, were simultaneously suffering through the impacts of serious drought with important consequences for prices of key food commodities such as corn, wheat, and soybeans. The city of Duluth, Minnesota, was devastated by record flash floods that tore through the city in late June following 7 inches of rain that fell on ground already saturated as a result of exceptionally heavy earlier precipitation. Less than a week later, Tropical Storm Debby stalled for 2 days over the Gulf of Mexico and dropped more than 2 feet of rain on areas of northern Florida, causing the Anclote River to rise by more than 27 feet. The winter of 2011–2012 was extremely dry in Colorado with precipitation averaging less than 15% of normal. High temperatures and exceptionally low relative humidity set the stage for a major fire season in the early summer of 2012. Particularly damaging was the fire that broke out in Waldo Canyon, 10 miles northwest of Colorado Springs, on June 23. By the time it was finally brought under control, this fire had destroyed a record number of homes in the Colorado Springs region, forcing more than 32,000 residents, including personnel from the US Air Force Academy, to evacuate. At about the same time, Russia was forced to declare a state of emergency to deal with out-of-control fires in the Khabarovsk Territory, while China was suffering the effects of torrential rain that fell across the central and southeastern part of the country, destroying crops over more than 700,000 hectares with a cost in excess of $1.6 billion.

There are also unintended consequences to all of this unusual weather. High temperatures over a large region of the central and eastern United States during the summer of 2012 led to an important increase in demand for electricity for air conditioning. Despite the increase in demand, production of electricity from nuclear power plants fell to the lowest level experienced over the preceding 9 years: reactors from Ohio to Vermont were forced to limit operation due to restrictions occasioned by lack of access to water of acceptable temperature for cooling.[26]

KEY POINTS

1. *The greenhouse effect is real.* There is no credible controversy as to how this operates to give us a habitable planet. The prompt response to an increase in the concentration of a greenhouse gas such as CO_2 is a decrease in energy radiated to space, an increase in net energy absorbed from the sun.
2. *Concentrations of a number of key greenhouse gases are higher now than at any time over the past 800,000 years.* Again, there is no informed dispute on this. Measurements of gases trapped in polar ice provide unambiguous evidence in support of this conclusion.
3. *Global average surface temperatures have increased by about 1.1°C since 1890.* More than half of this increase took place since 1970. Reasons for the increase may be questioned, but there is no doubt that the increase is real.
4. *The ocean has been gaining heat on average over the past 40 years, most likely since the last quarter of the nineteenth century.* The increase in ocean heat content provides definitive evidence that the planet is currently out of equilibrium from an energy point of view, absorbing more energy from the sun than it is returning to space in the form of infrared radiation.
5. *Warming resulting from the increasing concentration of greenhouse gases has been offset until recently by cooling associated with aerosols formed as byproducts of conventional pollution.* The evidence supporting this assertion is generally indirect, having to do mostly with the fact that the changes in global temperatures as observed are less than would be expected based on the recorded increase in the concentration of greenhouse gases. There is an urgent need for focused studies to better define not only the direct but also the indirect effect of aerosols on the planet's energy budget and climate. Current understanding suggests that the aerosol offset will be less consequential in the future as countries adopt measures to minimize adverse effects of conventional pollution on human health and on the environment more generally. The pace of climate change is likely in this case to accelerate.
6. *Broad-scale impacts of climate change observed recently include expansion of the Hadley circulation regulating climatic conditions in the tropics and subtropics, for a decrease in the thickness of sea ice in the Arctic Ocean, for a decrease in the areal coverage of ice in this region in summer, and for a decrease in the mass of ice stored in the major ice sheets blanketing Greenland and Antarctica.* Expansion of the Hadley circulation, should it continue, may be expected to lead to an

extension of deserts to higher latitudes. Loss of ice cover in the Arctic is likely to trigger a series of self-reinforcing changes in high-latitude climate, resulting in further warming of the Arctic with additional loss of sea ice cover.
7. *As highlighted by Thomas Friedman, worldwide weather has become increasingly weird in recent years—more floods, more droughts, more violent storms, more extreme heat waves, and more out-of-control forest fires.* This pattern is likely to continue in the future, posing challenges for investments in infrastructure designed to cope solely with the extremes of weather observed in the past. The concern is that the historical record may not provide a reliable projection for the future. Distinguishing between extremes that can be assigned unambiguously to human-induced change as distinct from natural variability will pose a continuing challenge.
8. *The important societal implication of global warming is not that portions of the Earth are going to experience higher temperatures, increased precipitation, and increased droughts; it is that the extremes are going to become more extreme.* What was once a 1-in-100-year anomaly in the past, or even rarer, may develop in the future as a 1-in-10-year event or even worse, threatening the viability of expensive investments in infrastructure ranging from roads, to bridges, to transportation systems, to industrial facilities, to residences, and, among other challenges, to the orderly function of systems critical for the supply of essential energy and communications services.
9. *Sea level has risen by about 6 cm over the 19 years for which precise data are available from space-based measurements.* The rate of increase is greater than was previously predicted. A key question for the future concerns the fate of the major ice sheets on Greenland and Antarctica. Data from the GRACE (Gravity Recovery and Climate Experiment) satellite provide definitive evidence for significant, and accelerating, net loss of ice from both of these regions over the period April 2002 to February 2009. A major release of ice from either of these environments in the future could pose serious problems for the security of coastal communities worldwide. Given the surprisingly extensive surface melting observed during early July 2012 in Greenland, it is clearly important that these systems be subjected to continuing scrutiny.

Notes

1. Energy from the sun originates from a region of the solar atmosphere where the temperature is about 5,600°C. The higher the temperature of a body, the shorter the wavelength of the light it emits. Given its high temperature, the sun emits radiation primarily in the visible region of the spectrum. The Earth, in contrast, is relatively cold (as least compared to the sun). It emits radiation at longer wavelengths, in the infrared region of the spectrum. Were you to take a spectrum of the Earth from space on the day side of the planet, you would observe that the spectrum exhibits two peaks, one in the visible corresponding to reflected sunlight, the second in the

infrared corresponding to radiation produced by the Earth. The two peaks would be relatively well separated. It is realistic therefore to talk about sunlight and planetary (infrared) light as distinct systems. This is the basis for the presentation in Figure 4.1, where the fate of visible sunlight is displayed on the left with the fate of infrared radiation indicated on the right.

2. The reflectivity of a specific surface depends not only on the nature of the surface but also on the viewing geometry (the difference between the angle of incidence of the sunlight and the direction from which the reflected light is observed). The albedo of fresh snow can be as high as 85% with forests typically as low as 5%–10%. The albedo of desert regions averages about 25%. As indicated, the global average albedo is a little more than 31% with an important contribution due to reflection from clouds and brightly colored particles suspended in the atmosphere (the latter, as indicated earlier, referred to as aerosols).

3. A great deal of energy is required to change the phase of water, to convert water from liquid to vapor. Evaporation of water from the ocean serves to cool the ocean. When the vapor condenses subsequently in the atmosphere and falls out as rain or snow, this energy is released in the atmosphere. The ocean is cooled by evaporation; the atmosphere is heated by precipitation. The evaporation/precipitation cycle provides an important means for transfer of energy from the ocean to the atmosphere with additional transfer associated with the evaporation/precipitation cycle on land. It takes 2,400 BTU to change the phase of 1 kg of water from liquid to gas. The total mass of water involved globally in the annual evaporation/precipitation cycle amounts to about 6×10^{17} kg or 1,170 kg (1.17 tons) per square meter averaged over the entire surface of the Earth. The energy deployed annually to cycle this water amounts to about 1.3 million quad, greater than the total energy consumed on a commercial basis globally by almost a factor of 3,000.

4. Emission of radiation depends on the fourth power of the temperature. A portion of the radiation emitted from the Earth's surface, 10% as indicated by Figure 4.1, escapes directly to space. The balance is absorbed, mainly by water vapor, in the atmosphere only a short distance above the surface where the temperature is only a little less than that of the surface. This accounts for the fact that the return of radiation from the atmosphere to the surface differs only marginally from the radiative energy the atmosphere receives from the surface, by less than 7.5%. Hence, the atmosphere is said to be optically thick with respect to the bulk of the infrared radiation emitted by the atmosphere (and the Earth's surface). What that means is that photons (the basic components of radiation) are only able to move a very short distance before being absorbed and reemitted by the atmosphere.

5. The frozen-over Earth is not simply a figment of the imagination. There were at least four occasions during the Neoproterozoic Era between 750 and 580 million years ago when the Earth was totally frozen over, a condition referred to by California Institute of Technology geologist Joseph Kirschvink (1992) as the Snowball Earth. Hoffman et al. (1998) attributed the Snowball Earth condition to a significant decrease in the concentration of the key greenhouse gas CO_2 triggered by a precipitous decline in tectonic activity (cf. note 4, Chapter 2, for the importance of tectonic activity in cycling carbon between sediments and the atmosphere).

6. The decrease in the flux of radiation to space occasioned by an instantaneous increase in the concentration of CO_2 can be estimated with precision based on well-defined properties of the molecule.

7. The greater the climate sensitivity, the greater the energy input required to reach the new equilibrium, and as a consequence the longer the time required to provide the necessary input.

8. The climate of the past 800,000 years was marked by prolonged periods of extremely cold conditions (ice ages) punctuated by relatively brief periods of warmth (interglacials). Sea level was about 110 m lower during the last ice age, which ended approximately 20,000 years ago. Water withdrawn from the ocean during the ice ages was sequestered in vast continental ice sheets. The water removed from the ocean was typically isotopically light compared to the water left behind in the ocean: the abundance of water that remained in the ocean was enriched in the heavier forms of hydrogen (deuterium, D) and oxygen (^{18}O). The isotopic composition of the shells of organisms preserved in sediments provides a valuable record of the changing isotopic composition of the ocean and consequently a surrogate for the changing mass of water preserved in the continental ice sheets and thus for global climate.

9. The concept of radiative forcing as employed by IPCC and as used commonly in a variety of climate studies is relatively simple. It seeks to provide a rough estimate of the change in energy flux at the top of the atmosphere in response to an assumed change in atmospheric composition. In the case of a change in the composition of a greenhouse gas such as CO_2, the relative abundance of which is assumed to be constant as a function of altitude, it first calculates the change in the temperature of the stratosphere subject to an assumption that the temperature everywhere in the stratosphere adjusts to the new composition according to what is referred to as local radiative equilibrium. That is to say, at every level in the stratosphere and above, it assumes that there should be no net source of energy: sources and sinks of energy are taken to be in precise balance. The temperature of the lower atmosphere (the troposphere) and surface are fixed at levels that prevailed prior to the postulated change in composition. The resulting change in the flux of energy at the base of the stratosphere (the tropopause) is then calculated. Since there is no net source or sink for energy above the tropopause, the change in energy at the tropopause should be the same as the change in energy at the top of the atmosphere, an indication therefore of the resulting imbalance in global energy. The justification for this procedure (why temperatures in the stratosphere should be allowed to adjust while temperatures in the troposphere and at the surface are fixed) has to do with the expectation that while the stratosphere (closer to space) can adjust relatively rapidly to the compositional change, the temperature in the troposphere and at the surface will be controlled to a larger extent by the dynamics of the climate system, by prevailing conditions at the surface, and by the vagaries of the changing circulations of the atmosphere and ocean in response to the changing energetics of the global system. This adjustment can take years or even hundreds of years. The data for radiative forcing attributed to the changing concentrations of CO_2, CH_4, and N_2O as presented in Figure 4.3 were computed following these procedures. It was assumed that the changes in composition implied by the surface measurements as indicated in the figure applied at all altitudes (that the gases were well mixed). In practice this assumption is reasonable for CO_2 and N_2O, which are relatively long lived. It is more questionable for CH_4, the lifetime in the atmosphere for which is estimated at a little less than 10 years.

10. This estimate takes into account not only the changes observed in the concentrations of CO_2, CH_4, and N_2O, it allows also for contributions from a number of other greenhouse agents, including sulfur hexafluoride (SF_6) and a variety of species, referred to collectively as chlorofluorocarbons or CFCs implicated earlier in the loss of ozone in the stratosphere not only at mid-latitudes but even more spectacularly in polar regions (as exemplified by the so-called hole in the ozone layer over Antarctica reported first in 1985 and a companion hole observed more recently over the Arctic). The nations of the world responded aggressively in Montreal in 1987 to

curtail and eventually eliminate most of the emissions of the major ozone-depleting CFCs. The Montreal Protocol, established at that meeting, has been generally successful and the contribution of CFCs to radiative forcing, as indicated in Figure 4.4, has dropped significantly over the past several decades. The estimate by Hansen et al. (2011) for what they define as effective radiative forcing in 2003 takes into account not only the primary effects of the key greenhouse gases but also secondary effects, notably the increase in the abundance of H_2O in the stratosphere and the increase in the abundance of ozone in the troposphere estimated to result from the breakdown of CH_4 (ozone is also a greenhouse gas).

11. Plant material ingested by cattle, sheep, goats, buffalo, and deer is stored temporarily in the first compartment of their stomachs, known as the rumen, where it is transformed to more digestible organic matter by a complex population of bacteria that thrive in this environment (the animals that rely on this means for food processing are known, for obvious reasons, as ruminants). To the extent that humans have come to rely increasingly on animal intermediaries for food, and the population of the more important ruminants is increasing, it is clear that animal husbandry must be recognized as an important anthropogenic (human-related) source of methane. A typical domestic cow produces as much as 200 liters of methane per day, emitted through a combination of belching and flatulence. It is not surprising that the ruminant source should be significant: there are more than a billion cattle in the world today.

12. Methane is removed from the atmosphere primarily by reaction with the hydroxyl radical OH. Production of OH is initiated by absorption of ultraviolet sunlight by ozone (O_3). Its concentration is determined by a complex suite of chemical reactions involving not only methane (CH_4) and O_3 but also oxides of nitrogen (NO_x), carbon monoxide (CO), and water vapor (H_2O). It is sensitive also to the abundance of O_3 in the stratosphere, which ultimately determines the flux of ultraviolet radiation reaching the lower regions of the atmosphere, where the loss process is primarily concentrated. The lifetime of CH_4 in the atmosphere is estimated at 9.2 years, in contrast to the much longer lifetime, hundreds of years, associated with CO_2. For a more detailed discussion of the chemical processes impacting not only CH_4 and CO_2 but also N_2O, see McElroy (2002).

13. Nitrogen is an essential component of all living organisms. The bulk of the Earth's nitrogen is present in the atmosphere in the chemically inert form of N_2. In this form the element is essentially inaccessible to living organisms. To be made available, the strong bond than binds the N atoms in N_2 must be sundered—we say that the nitrogen must be fixed. Nitrogen fertilizer is produced by subjecting N_2 to extremely high temperatures, most commonly using natural gas as the intermediate energy source. By producing fertilizer and by cultivating certain crops with the capacity to fix nitrogen (plants in the legume family that includes, for example, soybeans), it is clear that humans are now playing an important, arguably dominant, role not only in the global carbon cycle but also in the global nitrogen cycle.

14. Clouds form in the atmosphere when temperatures are low enough and concentrations of water vapor are high enough to prompt a phase change, production of either liquid water or ice. This phase change does not generally occur spontaneously; it proceeds normally on the surfaces of preexisting particulate matter with an affinity for water, what atmospheric scientists refer to as condensation nuclei. The greater the number of sites available for condensation, the greater the number of cloud particles that form and thus the smaller their individual size (the available source of water is distributed over more individual sites). An increase in the number of cloud particles leads to an increase in reflection of sunlight from the cloud.

15. Regions of the world where practices favor burning of biomass as a means to clear land and where crop and animal residues provide a source of cheap energy offer a potential localized exception to this general conclusion.

16. For a more detailed, though accessible, account of the complex influence of the ocean on climate, the reader is referred to Vallis (2011).

17. During the cold La Nina phase, waters in the surface tropical Pacific are driven across the ocean from east to west under the influence of the trade winds. Water leaving the eastern region of the ocean off the coast of South America is replaced by cold nutrient-rich waters upwelling from below, spawning one of the world's richest fisheries. The westward transport of water cannot, however, continue indefinitely. Water will tend to pile up in the western portion of the basin. Eventually the head created by this pile-up triggers an adjustment in which warm western water flows back across the ocean eventually to the east, choking off the supply of cold water and nutrients that previously fed the rich fishery in this region. The trade winds slacken, or even reverse direction, as the spread of this warm water contributes to a more uniform thermal environment across the ocean. The region of intense precipitation shifts from the western to the central region of the ocean, contributing to a drought in the previously water-rich regions of Indonesia and surrounding countries, with intense rainfall in the east, notably over Peru and Ecuador. This defines the warm or El Nino phase of the oscillation. The circulation of the atmosphere adjusts to the changing distribution of temperatures in the Pacific, most notably to the eastward shift in the region of intense convection and associated precipitation and heating of the atmosphere, contributing to the widespread changes observed in global weather systems. The El Nino condition, as indicated, is accompanied by intense precipitation over coastal South America usually over the Christmas season, hence the name associated with this condition, Spanish for "little boy," referring to the Christ child. La Nina or "girl child" was an obvious choice of terminology for the opposite phase of the ENSO phenomenon.

18. Richard Muller is a physicist at the University of California at Berkeley who advertises himself as a "converted skeptic" on the issue of human-induced climate change. Quoting from the op-ed article he published in *The New York Times* on July 28, 2012, he stated: "Three years ago I identified problems in previous climate studies that, in my mind, threw doubt on the very existence of global warming. Last year, following an intensive research effort involving a dozen scientists, I concluded that global warming was real and that prior estimates of the rate of warming were correct. I am now going a step further: humans are almost entirely the cause." The community of climate scientists who devoted their intellectual energies to the subject for many more years than Muller may have been excused a reaction of "we told you so." From a political viewpoint, though, Muller's announcement was significant in confronting the views of some public skeptics on the issue of human-induced climate change. His approach, however, has been somewhat unorthodox in that at the time he published his article in *The New York Times* and was interviewed on a number of national US television networks, his scientific papers had not been as yet subjected to the traditional review by peers and had not been published in any traditional professional scientific journals. Despite these reservations, having studied Muller and his colleagues' papers as made available on his website, I believe that they make an important contribution to our understanding of human-induced climate change.

19. Booth et al. (2012) argue that changing patterns of sulfur emissions (including emissions associated with volcanoes) may have played a role in determining the variability of the AMO at least over the past century. The persistence of the variability observed over the longer

interval covered by Knudsen et al. (2012) would appear to argue in favor of the ocean connection, although arguably both influences may have played a role over the more recent period.

20. The actual increase is probably greater than implied by the data in Figure 4.12, given the limitations in coverage available from the ARGO program at high latitudes (in regions of the ocean subject to seasonal ice cover) and the comparative absence of data for depths greater than about 1750 m.

21. Output of energy from the sun varies over a range of approximately 0.25 W m^{-2} between solar maximum and solar minimum. Output peaked most recently in about 2001. The recent minimum has been unusually extended, lasting from about 2005 to about 2011. Solar activity is now again on the rise. Energy output from the sun varies on a characteristic period of about 11 years.

22. HMS Challenger conducted the first essentially global-scale study of the world's oceans between 1872 and 1876 covering a track of approximately 69,000 nautical miles, extending over large regions of the Atlantic and Pacific Oceans, both north and south, with more limited sampling of the Indian Ocean, mainly in this case at higher southern latitudes. Roemmich et al. (2012) argued that the changes in ocean heat content inferred from comparison of the Challenger data with the more recent measurements from ARGO represent most likely a lower bound to the actual changes that developed over the intervening time interval.

23. We might expect a modest increase in evaporation associated with increased input of energy to the ocean. This increase is unlikely, however, to alter the basic conclusion drawn here: that we should expect greater extremes in precipitation both on the dry and wet side. When and where it rains, and where it does not, is determined ultimately by the details of the atmospheric circulation. Episodes of intense precipitation are likely to be associated primarily with weather systems of marine origin. Episodes of extreme drought are most likely to be experienced in continental interiors. As soils dry out in these environments, temperatures will tend to rise in the face of a reduction in the cooling impact of evaporation, posing problems for agriculture in some regions while increasing the risk of fires in others. As discussed later, this is precisely what has been happening in the United States over the past several years, and the problems have not been confined to the United States.

24. The Northwest Passage refers to a sea route along the northern coast of North America passing through the Bering Straits traversing the Chukchi and Beaufort Seas. When open to shipping, as it has been for a few months over the past several years, the Northwest Passage offers the opportunity to significantly reduce distances and times for shipping of goods between Asia and northern Europe. Canada claims territorial rights to the Passage. These claims are disputed, however, by other nations with interests in the Arctic.

25. In a steady state, gains and losses of ice are expected to be in balance. Gains of water from precipitation would be offset by comparable losses due to release of ice to the ocean. An increase in surface melting over and above the mean could result, however, in increased transfer of water to the ocean, either in the form of liquid reaching the base of the ice sheet or as a result of instabilities in the ice sheet triggered by freezing of liquid in cracks of the ice sheet as noted here.

26. Water, after it is used for cooling, is normally returned to the environment from which it was derived in the first place, a river for example. Access to cooling water is customarily restricted if the temperature at which the water is returned to the environment is judged to be too high, likely to have a negative impact on the environment.

References

Booth, B. B. B., N. J. Dunstone, P. R. Halloran, T. Andrews, and N. Bellouin. 2012. Aerosols implicated as a prime driver of twentieth-century North Atlantic climate variability. *Nature* 484, no. 7393: 228–232.

Charney, J. G., A. Arakawa, D. Baker, B. Bolin, R. Dickinson, R. Goody, C. Leith, H. Stommel, and C. Wunsch. 1979. *Carbon dioxide and climate: A scientific assessment*. Washington, DC: National Academy of Science Press.

Emanuel, K. 2007. Environmental factors affecting tropical cyclone power dissipation. *Journal of Climate* 20, no. 22: 5497–5509.

Fu, Q., C. M. Johanson, J. M. Wallace, and T. Reichler. 2006. Enhanced mid-latitude tropospheric warming in satellite measurements. *Science* 312, no. 5777: 1179. doi: 10.1126/science.1125566.

Hansen, J., and M. Sato. 2012. *Paleoclimate implications for human-made climate change in climate change: Inferences from Paleoclimate and regional aspects*, ed. A. Berger, F. Mesinger, and D. Sijacki. New York: Springer.

Hansen J., M. Sato, P. Kharecha, and K. von Schuckmann. 2011. Earth's energy imbalance and implications. *Atmospheric Chemistry and Physics* 11: 13421–13449.

Hansen, J., P. Kharecha, and M. Sato. 2013. Climate forcing growth rates: Doubling down on our Faustian bargain. *Environmental Research Letters* 8, no. 1 (March): 1–9.

Hoffman, P. F., A. J. Kaufman, G. P. Halverson, and D. P. Schrag. 1998. A Neoproterozoic Snowball Earth. *Science* 281, no. 5381: 1342–1346.

Huang, J. L., and M. B. McElroy. 2012. The contemporary and historical budget of atmospheric CO_2. *Canadian Journal of Physics* 90, no. 8: 707–716.

Intergovernmental Panel on Climate Change (IPCC). 2007. *Climate change 2007: The physical science basis*, edited by S. Solomon, Q. Dahe, M. Manning, Z. Chen, M. Marquis, K. B. Averyt, M. Tignor, and H. L. Miller. Cambridge: Cambridge University Press.

Kirschvink, J. L. 1992. Late Proterozoic low-latitude global glaciation: The Snowball Earth. In *The Proterozoic biosphere: A multidisciplinary study*, ed. J. W. Schopf and C. C. Klein, 51–52. Cambridge: Cambridge University Press.

Knudsen, M. F., M. S. Seidenkrantz, B. Holm Jacobsen, and A. Kuijpers. 2011. Tracking the Atlantic multidecadal oscillation through the last 8,000 years. *Nature Communication* 2, no. 178, doi:10.1038/ncomms1186.

Levitus, S., J. I. Antonov, T. P. Boyer, O. K. Baranov, H. E. Garcia, R. A. Locarnini, A. V. Mishonov, J. R. Reagan, D. Seidov, E. S. Yarosh, and M. M. Zweng. 2012. World ocean heat content and thermosteric sea level change (0–2000 m), 1955–2010. *Geophysicals Research Letters* 39, no. L10603, doi:10.1029/2012GL051106.

Lu, X. M. B. McElroy, G. Wu, and C. P. Nielsen. 2012. Accelerated reduction in SO_2 emissions from the US power sector triggered by changing prices of natural gas. *Environmental Science and Technology* 46, no. 14: 7882–7889.

McElroy, M. B. 2002. *The atmospheric environment: Effects of human activity*. Princeton, NJ: Princeton University Press.

Muller, R. A., J. Curry, D. Groom, R. Jacobsen, S. Perlmutter, R. Rohde, A. Rosenfeld, C. Wickham, and J. Wurtele. 2012. *Decadal variations in the global atmospheric land temperatures*. Berkeley: University of California Press.

Murphy, D. M., S. Solomon, R. W. Portmann, K. H. Rosenlof, P. M. Forster, and T. Wong. 2009. An observationally based energy balance for the Earth since 1950. *Journal of Geophysical Research* 114, no. D17107, doi:10.1029/2009JD012105.

Roemmich, D., W. J. Gould, and J. Gilson. 2012. 135 years of global ocean warming between the Challenger expedition and the Argo Programme. *Nature Climate Change* 2, no. 6: 425–428.

Seidel, D. J., and W. J. Randel. 2007. Recent widening of the tropical belt: Evidence from tropopause observations. *Journal of Geophysical Research* 112, no. D20113, doi: 10.1029/2007JD008861.

Smith, S. J., J. van Aardenne, Z. Klimont, R. J. Andres, A. Volke, and S. D. Arias. 2011. Anthropogenic sulfur dioxide emissions: 1850–2005. *Atmospheric Chemistry and Physics* 11, no. 3: 1101–1116.

Vallis, G. K. K. 2011. Climate and the oceans. Princeton, NJ: Princeton University Press.

Velicogna, I. 2009. Increasing rates of ice mass loss from the Greenland and Antarctic ice sheets revealed by GRACE. *Geophysical Research Letters* 36, no. L19503, doi: 10.1029/2009GL04022.

5

Human-Induced Climate Change

ARGUMENTS OFFERED BY THOSE WHO DISSENT

CHAPTER 4 PRESENTED an extensive account of current understanding of climate change. The evidence that humans are having an important impact on the global climate system is scientifically compelling. And yet there are those who disagree and refuse to accept the evidence. Some of the dissent is based on a visceral feeling that the world is too big for humans to have the capacity to change it. Some is grounded, I believe, on ideology, on an instinctive distrust of science combined with a suspicion of government, amplified by a feeling that those in authority are trying to use the issue to advance some other agenda, to increase taxes, for example. More insidious are dissenting views expressed by scientists on the opinion pages of influential newspapers such as *The Wall Street Journal* (*WSJ*). If scientists disagree, the implication for the public is that there is no urgency: we can afford to wait until the dust settles before deciding to take action—or not, as the case may be. Missing in the discourse triggered by these communications is the fact that, with few exceptions, the authors of these articles are not well informed on climate science. To put it bluntly, their views reflect personal opinion and in some cases explicit prejudice rather than objective analysis. Their communications are influential, nonetheless, and demand a response.

I begin by addressing some of the general sentiments expressed by those who are either on the fence as to the significance of human-induced climate change or who may already have made up their minds that the issue is part of an elaborate hoax to mislead the public. There are a number of recurrent themes:

1. The data purporting to show that the world is warming have been manipulated by climate scientists to enhance their funding or for other self-serving reasons.
2. Climate science is complicated; scientists cannot predict the weather. Why should we believe that they could tell us what is going to happen a decade or more in the future?
3. The planet has been warmer in the past; we survived and maybe even prospered.
4. The recent change is part of a natural cycle—nothing to do with the impact of humans.
5. A little warming could be beneficial.

Following discussion of these issues, I will address a number of the more specific technical points highlighted in *WSJ* opinion articles.

GENERAL CLAIM NUMBER 1

The data have been manipulated. This is an assertion that goes back to the selective release of emails hacked illegally in November 2009 from a server hosted at the Climate Research Unit (CRU) of the University of East Anglia in the United Kingdom—a development that came to be known subsequently as Climategate. The CRU group had long been active, together with others, in assembling the record of changes in globally averaged surface temperature. Critics suggested that the emails indicated a pattern of deception, including efforts to manipulate data and to suppress views of scientists whose conclusions differed from those of the CRU. A number of committees both governmental and private, not only in the United Kingdom but also in the United States, investigated the allegations and found no evidence for scientific misconduct.

It should be emphasized that the CRU group did not have a monopoly on studies of globally averaged surface temperature. The data presented in Figure 4.5, for example, draw on results from an unrelated study at the NASA Goddard institute for Space Studies (GISS) in New York. Furthermore, as discussed in Chapter 4, a totally independent investigation by Richard Muller and colleagues at the University of California at Berkeley reached conclusions similar to those reported by GISS and CRU. As indicated, Muller began as a skeptic expecting to identify errors in the published analyses. The fact that he concluded otherwise speaks to the credibility of the independent temperature analyses. The scientific method, where data were freely available and could be independently assessed, has worked well in this case. The data were not manipulated. Results from a variety of analyses confirm that globally averaged surface temperatures have increased by about 1°C over the past 130 years.

GENERAL CLAIM NUMBER 2

Climate models are complicated and results cannot be trusted. Let's state the obvious from the outset: models used to simulate climate are indeed complicated. To do the job, they must provide a realistic description of the atmosphere, ocean, biosphere, and cryosphere (the ice world) as a coupled dynamical system. Time scales associated with important phenomena range from hours to years to even hundreds of years. Spatial scales of consequence are similarly wide ranging—centimeters to meters to hundreds of kilometers. Judicious choices must be made in selecting the spatial and temporal resolution of the models. Many of the key processes—the formation of clouds, for example—must be parameterized, that is to say they must be treated not from first principles but on the basis of empirically derived, physically rationalized associations inferred from coarser scale resolution properties of the models. The models are exceptionally demanding, requiring major commitments of computer resources, straining the power and capability of current computer technologies and resources. Most of the world's frontier models are housed either in government laboratories or in national facilities, supported by large teams of disciplinary professionals—both programmers and scientists. This is big science, implemented accordingly.

State-of-the-art simulations are tuned typically to reproduce past variations of global average surface temperature. Early studies restricted attention to the impact of the changing concentration of greenhouse gases and had difficulty in reproducing the trend observed over the past 130 years. Typically, they predicted an increase in global average temperature *greater* than what was observed. This was followed by the recognition, as discussed in Chapter 4, that aerosols could provide a source of negative radiative forcing, offsetting the positive forcing contributed by greenhouse gases. Given the uncertainty in specifying the magnitude and time history of the role of aerosols in overall radiative forcing, it is clear that the introduction of a potential aerosol influence of uncertain magnitude and sign provided models with an important additional degree of freedom, thus facilitating efforts to fit the historical global surface temperature data.

Close inspection of a number of the models employed by the IPCC indicates that while the models are generally successful in reproducing trends in global average surface temperature, details of the climates simulated by them can vary significantly from model to model. Differences include variations in representations of regional climates, in addition to significant differences in treatments of the hydrological cycle—greater or lesser rainfall, greater or lesser cloud cover, differences in cloud altitudes and thicknesses, and so on. Under these circumstances, the ability of models to project future conditions is questionable. There is a view, though, that useful information can be obtained by combining results from a range of models (an ensemble). The argument is that errors in individual models may be compensated in this case. Results from a combination of models, it is suggested, may be more credible, more robust, than those from any single simulation. Use of model ensembles is common in predicting, days in advance, short-term variations in weather—the path of hurricanes such as Sandy, for example. The hope is that they may be similarly instructive in forecasting general features of potential future climates.

It should be pointed out that in making the case in Chapter 4 as to why one should take the prospect for human-induced climate change seriously, I did not explicitly appeal to results from these complicated computer models. I chose rather to focus on the fact that given the inexorable rise in the concentration of greenhouse gases, for which we humans are unquestionably responsible, the planet is likely to get warmer, probably much warmer, in the future, continuing the trend observed over much of the past 130 years. A warmer atmosphere, I argued, is likely to hold more water vapor. This implies that when it rains, it is likely to rain more than it would otherwise. If precipitation on a global scale remains relatively constant, this would imply that heavy rain in a particular region at a particular time should be compensated by less rain elsewhere—more floods, more droughts. Drought combined with high temperatures will increase the likelihood of serious wildfires. The key point is that weather experienced in the past is unlikely to provide a useful guide to what we may expect in the future. We don't need computer models to alert us to the danger. We need only look to the especially unusual weather we have experienced over the past several years.

GENERAL CLAIM NUMBER 3

The planet was warmer in the past. Chapter 1 included a brief overview of past climates. The Earth was generally warm, much warmer than it is today, or prospectively in the immediate future, during the geological period referred to as the Cretaceous, which lasted from 145 to 66 million years before present (BP). Warm conditions persisted up to a few million years ago, with palm trees growing in Central Asia as recently as 5 million years BP. All this, of course, was before humans made their entry onto the stage of life. For much of the time that humans have been around, the planet has been cold, transitioning between prolonged exposure to ice ages punctuated by relatively brief interludes of modest interglacial warmth.

The world exited the last ice age about 20,000 years ago. In the normal course of events, the planet would have transitioned to maximum warmth approximately 7,000 to 5,000 years ago. The planet has cooled moderately since then. There were intervening times when it was relatively warm and times when it was comparatively cold. The medieval period, from about AD 950 to 1200, for example, was an interlude when it was generally warm. Climate scientists refer to this interval as the Medieval Optimum. The period from AD 1400 to 1850, referred to as the Little Ice Age, was unusually cold. I suggested in Chapter 1 that we might now be well on our way to the next Big Ice Age were it not for the buildup of greenhouse gases that has resulted from our increased use of fossil fuels over the past several centuries. But I also suggested that we might have already gone too far!

The best evidence suggests that the Earth is warmer now than it has been at any time over the past several thousand years: we probably surpassed the relative warmth of the Medieval Optimal at some point in the middle of the last century. Living as we do today in concentrated (generally urban) communities, we are increasingly vulnerable to extremes of weather and unanticipated changes in climate. The population of the planet during the medieval period was about 300 million; today it exceeds 7 billion on a path to more than 9 billion. Our contemporary lifestyles depend critically on the infrastructure we have developed at significant expense to ensure reliable access to necessary sources of food, shelter, and energy. The central problem is not simply that the world is getting warmer. It is that the expression of weather on a regional basis is changing. The fundamental challenge we face is that the infrastructure—buildings, roads, bridges, ports, power sources, and so forth—on which we have come to rely over the past several decades may be ill equipped to deal with the anticipated new climate normal.

GENERAL CLAIM NUMBER 4

The climate change now underway is part of a natural cycle. This is a view inspired by the knowledge that the medieval period was a time of relative warmth, succeeded by a time when the world was relatively cold. If it is warmer now than it was a few hundred years ago, why not attribute the recent warmth to a natural cycle, a change for which we bear no responsibility?

As discussed earlier, one would expect in the normal course of events that global climate, responding to the changing orbital parameters of the planet, should have been on a general cooling path over the past several thousand years. The Medieval Optimum and the Little Ice Age could be interpreted as natural fluctuations on a multicentury time scale, warm and cold respectively, punctuating this long-term trend. These fluctuations could relate to changes in the circulation of waters in the deep ocean, the so-called Conveyor Belt.[1] We might expect the Little Ice Age to be followed in this case by a century or so of relative warmth. This would be superimposed, however, on a continuing long-term cooling trend. The problem with this suggestion is that it ignores the fact that the concentration of greenhouse gases is much higher now than it was during the putative earlier cycles, indeed higher than at any time over the past 800,000 years. As discussed in the preceding chapter, there is indisputable evidence for a net input of energy to the global climate system: The planet as a whole is gaining energy. The cooling trend that defined conditions over the past several thousand years has been supplanted now by warming. While natural fluctuations may still play a role, there can be little doubt that the warming observed over the past 130 years is real and significant and that human activity is largely responsible. Failing precipitous action to address this disturbance, the weight of the evidence suggests that the current trend will continue and indeed that the pace of change is likely to quicken in the future.

GENERAL CLAIM NUMBER 5

The benefits from a warming world could be positive. The winter of 2012 was unusually mild in the northeastern United States. We had little snow, temperatures as much as 5 degrees above normal, shirtsleeves weather in March, and bills for heating a fraction of what they were the previous winter. It would have been difficult to find anyone in my community who would have objected to this pleasant experience—what some described as the year without a winter. Switch forward to the summer of 2012. You would certainly get a very different reaction in this case from the farmer in Iowa who watched his corn crop wither in the field in the face of record heat and drought, or from the rancher in west Texas who was forced to cull his cattle herd when confronted with an inadequate supply of grass for grazing. A little warming in winter we can certainly live with, although there might be subtle and undesirable implications for natural ecosystems (disruptive changes in the normal rhythm of the seasons, for example). Record high temperatures in summer, particularly in regions that are already pretty warm, pose a very different problem—the likelihood of a decrease in agricultural productivity, the potential for serious illness or even death from heat stroke, increased bills for air conditioning, the risk for disruption of power supplies, and a host of other challenges. The fact is that we are seeing more incidences of record high temperatures around the world in recent years and fewer incidences of record lows. The overall increase in surface temperatures is not an unqualified positive. The ultimate challenge, though, may have to do with our ability to respond to changes in climate for which we may be singularly ill prepared.

I turn now to a discussion of issues raised in the articles that appeared in *The Wall Street Journal* in early 2012. There were 16 authors from a variety of disparate scientific fields who lent their names to the first of these publications (Allegre et al. 2012a). This appeared on January 12, with a follow-up (Allegre et al. 2012b) on February 21. The third article (Happer 2012), which appeared on March 27, was authored by Princeton optical physicist William Happer, a coauthor on the earlier articles and a long-term climate skeptic. In addition to his post at Princeton, Happer serves as chairman of the board of the conservative Marshall Institute, which is based in Arlington, Virginia.

I propose to treat the issues raised in these articles sequentially, introducing particular individual criticisms with direct quotes. The Allegre et al. (2012a) article begins by offering advice to hypothetical candidates for public office as follows: *"A large and growing number of distinguished scientists and engineers do not agree that drastic actions on global warming are needed."* It continues with a statement that *"the reason [these distinguished scientists and engineers take this position] is a collection of stubborn scientific facts."* What precisely are the facts with which these writers take issue?

WSJ CRITICISM NUMBER 1

"The most inconvenient fact is the lack of global warming for well over 10 years" (Allegre et al. 2012a). The challenge here has to do with interpretation of the record of changes in surface temperature, as summarized for example in Figures 4.5, 4.10, 4.11, and 4.12. The results displayed in Figure 4.5 reproduce annual mean and 5-year running averages for global surface temperatures dating back to 1880. The assertion is that the lack of evidence for warming over the most recent period poses a serious problem for those who would argue for a significant, human-induced change in global average surface temperature.

From even a cursory examination of the data in Figure 4.5, it is difficult to make the case that the recent trend in temperature should justify such a conclusion. The past decade, as pointed out by Trenberth et al. (2012), was in fact the warmest on record. Nine of the ten warmest years since 1880 occurred over the past decade with 2014 establishing a new 134-year record. Warming in the Arctic has been particularly impressive (Fig. 4.12). Based on the entire record, there is clear evidence for a long-term increase in surface temperatures, by about 1°C since 1880.

The trend over time is not monotonic (consistently in one direction). It varies from year to year and even decade to decade. There are times when temperatures leveled off, or even declined, for a decade or longer. It is important to recognize that surface temperatures provide at best an imperfect record of climate change. As discussed earlier, the heat capacity of the atmosphere is small compared to that of the ocean. To the extent that the Earth as a whole is gaining energy, we expect the bulk of the excess to be accommodated by the ocean. Evidence from the Argo float program (depths 0–2,000 m, Figure 4.13) clearly demonstrates that the ocean is warming. We noted the importance of the ENSO phenomenon. Surface temperatures may be expected to vary in response to ENSO-related changes in ocean circulation with related variations in ocean surface temperatures. The prevalence of the cold La Nina phase over the past decade provides at least a partial explanation for the recent stasis in the longer term upward trend of global average surface temperatures. Also relevant is the temporary decrease in energy output from the sun associated with the prolonged recent solar minimum (see Chapter 4, note 21).

Conclusion

Inferences drawn from the temperature record of the past decade have little to say about the future course of global climate. This particular criticism is without merit.

WSJ CRITICISM NUMBER 2

"*Computer models have greatly exaggerated how much warming additional CO_2 can cause*" (Allegre et al. 2012a). As discussed in the preceding chapter, the significance of any particular increase in the concentration of CO_2 can be measured in terms of the associated radiative forcing. Doubling of CO_2 with respect to preindustrial conditions is expected to result in an immediate decrease of about 4 W m^{-2} in infrared radiation emitted to space, with a comparable increase in net energy absorbed by the planet. This is fact, not conjecture: it depends simply on the properties of the molecule. The ocean absorbs a major fraction of this excess energy. Surface temperatures may be expected to increase, triggering additional changes in the radiative properties of the atmosphere, notably in the abundance of water vapor, the most important of the greenhouse gases, and in the prevalence and properties of clouds. Eventually, the climate will transition to a new equilibrium state in which energy in and energy out are again in balance. The final change in surface temperature is defined in terms of what is referred to as the climate sensitivity

(Chapter 4), which identifies the ultimate, equilibrium change in global average surface temperature projected to result from a doubling of CO_2.

The path to this new equilibrium depends on the details of the time history with which the excess heat is incorporated in the ocean. There is a consensus that 1 W m^{-2} of radiative forcing will result eventually in a rise in global average surface temperature by about 0.75°C, subject to an uncertainty (conservatively estimated) of about ±30%. This would imply that a doubling of CO_2 with respect to preindustrial conditions should lead to an increase in global average surface temperature by between 2°C and 4°C, with a central (most probable) value of 3°C (2.5°C–3.5°C with the same central value if we adopt the range of uncertainty quoted by Hansen and Sato as discussed in the preceding chapter). A change of this magnitude would clearly be significant. It should be noted that the sensitivity quoted here, 0.75°C/W m^{-2}, is supported by a variety of different lines of evidence, not simply by projections based on computer models. Relevant studies include analyses of the response to volcanic eruptions and to the changes in albedo and greenhouse gases associated with major glacial to interglacial transitions.

The view expressed by Allegre et al. (2012a), elaborated further by Happer (2012), that computer models exaggerate the impact of increasing levels of CO_2 reflects an opinion that the climate sensitivity simulated by computer models is too high. Let's concede, as we did earlier, that there are important uncertainties associated with complicated computer models of climate. The assertion by Allegre et al. (2012a) and Happer (2012) is that the models overestimate the significance of the warming induced by rising levels of CO_2.

Allegre et al. (2012a) and Happer (2012) question specifically the importance of the amplifying feedbacks incorporated in the models, notably the projected increase in the concentration of water vapor. The increase in global average temperature predicted to result from a doubling of CO_2 could be reduced to about 1°C, they suggest, if current estimates of climate sensitivity were too high—by as much as a factor of 3. The observational data, specifically measurements from the Atmospheric Infrared Sounder Satellite (AIRS), however, tell a different story. They provide strong empirical support for the importance of the water vapor feedback and for the manner in which it is treated by climate models (Dessler et al. 2008; Dessler and Sherwood 2009). The contrary views expressed by Allegre et al. (2012a) and Happer (2012) are simply not supported by the facts.

It would be difficult in any event to account simultaneously both for the observed increase in surface temperatures and for the observed increase in ocean heat content if the climate sensitivity was as low as suggested by Allegre et al. (2012a) and Happer (2012). Were this the case, the increase in ocean heat content over the past several decades should have been significantly greater than what was actually observed (Fig. 4.13). It would be challenging further, if the climate sensitivity were this low, to account for the wide range of different climates (both hot and cold) observed in the past (as summarized in Chapter 1).

As indicated earlier in this chapter, climate models are complex. Their ability to provide detailed, high-resolution forecasts for conditions that might develop in any particular region of the world at any particular time in the future is limited. But, on the narrower question of whether current trends in emissions of CO_2 are likely to lead to a significantly warmer future world, however, there is little controversy. Present understanding of the sensitivity of the climate system to the global radiative (energy) imbalance that would result from a doubling of CO_2 would imply an increase in global average temperature of between 2.5°C and 3.5°C. To put this in context, a change of this magnitude would be comparable to the change that took place as the Earth emerged from the last ice age. Allegre et al. (2012a, 2012b) and Happer (2012) favor a much smaller value for climate sensitivity but fail to offer any convincing physical arguments to support their positions or to dispute the prevailing evidence to the contrary.

Conclusion

The assertion (Allegre et al. 2012a) that models "greatly exaggerate" the warming that could arise due to increasing levels of CO_2 is based on opinion not on careful analysis of the facts. The evidence points in a different direction. The criticism is without merit.

WSJ CRITICISM NUMBER 3

"Faced with this embarrassment [that warming has slowed down over the past decade] those promoting alarm have shifted their drumbeat from warming to weather extremes, to enable anything that happens in our chaotic climate to be ascribed to CO_2." (Allegre et al. 2012a). Happer (2012) makes a similar point: *"frustrated by the lack of computer-predicted warming over the past decade, some IPCC supporters have been claiming that 'extreme weather' has become more common because of CO_2."* We have already dealt with the question of the "embarrassment" resulting from the recent slowdown in warming: there is no reason for advocates of climate action to be "embarrassed" on this score. Imputing motives to climate scientists "shifting their drumbeat from warming" is polemical and unscientific. The question of whether we can expect a significant increase in the incidence of extreme weather as a result of human-induced changes in global climate, on the other hand, is serious and merits a reasoned response.

As discussed in Chapter 4, there is persuasive evidence that climate is changing. Warming is particularly rapid in the Arctic. As a result, the contrast in temperature between equator and pole is decreasing. The jet stream, the feature of high-speed air circling the globe separating temperate from polar regions, is slowing down as a consequence and is becoming notably less stable, subject to persistent meanders. The Hadley circulation, the dominant feature of the climate system in the tropics, is extending to higher latitudes, playing a role, it is argued, in the persistent drought observed most recently in the American Southwest and in the Mediterranean. The water vapor content of the atmosphere is increasing. As discussed earlier, this means that when it rains, it is likely to rain more than it did before. And, compensating, it will rain less elsewhere: the result—more floods, more droughts. Since condensation of water vapor provides the primary source of energy for storms, storms are expected to become more powerful.

And when ecosystems are subjected to an extended period of low moisture input, they are increasingly vulnerable to fire. Implication: more wildfires, strain to the breaking point on fire-fighting resources, lives lost, and extensive damage to property. These are facts supported by recent experience. Prospects for the future are daunting. Focusing on extreme weather is not an invidious strategy on the part of climate scientists to distract attention from problems with models. It defines a serious issue meriting serious attention. To suggest otherwise is disingenuous and unconstructive.

Conclusion

It is entirely appropriate for climate scientists to direct attention to the potential implications of "warming for extreme weather." The criticism that they do so for venal motives, to compensate for "embarrassment" over perceived deficiencies in their models, is without merit.

WSJ CRITICISM NUMBER 4

"The timing of the warming—much of it before CO_2 levels had increased appreciably—suggests that a substantial fraction of the warming is from natural causes that have nothing to do with mankind" (Allegre et al. 2012b). The global average temperature was relatively constant from 1880 to 1920. It increased by about 0.4°C between 1920 and 1940, with little change, or even a modest decline, between 1940 and 1975. It rose by about 0.6°C between 1975 and 2010. Happer's argument (2012), presumably, is that the rise in temperature (0.4°C) observed between 1920 and 1940 was comparable to that observed between 1975 and 2007 (0.6°C), despite the fact that the increase in CO_2 related to the earlier warming (7 ppm, 303 to 310 ppm) was much less than that associated with the latter (52 ppm, 330 to 382 ppm). The rationale for the conclusion he draws from this observation, that "a substantial fraction of the warming is from natural causes," is unsupported. In fact, there are plausible, persuasive, and indeed published alternative explanations that he chooses to ignore.

As discussed in Chapter 4, the evidence suggests that several factors have contributed to the change in climate observed over the past 130 years. There is agreement that the increase in the concentration of greenhouse gases has been primarily responsible for the warming observed over this interval. Warming from greenhouse gases has been offset, partially at least, by cooling due to reflective and hydroscopic aerosols. The hiatus in warming between 1940 and 1975 was attributed to the growth in emission of conventional sources of pollution, including notably sulfur produced by burning coal (Figs. 4.6 and 4.7). As indicated in Chapter 4, emissions of sulfur peaked in the early 1970s, reflecting primarily initiatives in the United States and Europe to deal with the perceived health impact of these emissions and to address the problem of acid rain. The decrease in emissions of sulfur and other sources of aerosol forming pollutants after 1975 provides a plausible explanation for the rapid increase in global average surface temperatures observed over the past several decades. As discussed in Chapter 4, warming is likely to increase in the future as China joins the United States and Europe in reducing its emissions of conventional pollutants. Absent the offsetting influence of aerosols, the pace of warming is projected to accelerate.

The atmosphere and ocean behave as a coupled system: that is to say, the atmosphere drives the ocean, and in turn the ocean drives the atmosphere. The ENSO phenomenon provides an excellent example of this coupling. In the cold (La Nina) phase, water is driven across the tropical Pacific Ocean from east to west by the northeasterly and southeasterly trade winds. Water in the surface zone leaving the eastern region of the ocean is replaced by cold water upwelling from below. As water at the surface moves across the ocean, it is warmed by absorption of sunlight. Eventually this water piles up in the west. When the level of the water in the west reaches a certain (critical) point, the system becomes unstable. The warm water in the west is driven back then across the ocean, blocking off the upwelling in the east, establishing an essentially uniform distribution of surface temperatures across the ocean. This dramatically alters the pattern of surface winds, setting up the opposite, warm (El Nino), phase of the cycle. Eventually, the trade winds recover, upwelling of cold deep water in the east resumes, surface waters are again driven to the west, and the cycle repeats. The ENSO phenomenon has an important influence on weather systems and climate over a wide region of the planet, as discussed in Chapter 4. Globally averaged surface temperatures are elevated by a few tenths of a degree during the warm phase of the cycle, depressed during the cold.

ENSO is but one of a number of quasi-cyclic phenomena contributing to what may be identified as natural sources of climate variability. Muller et al. (2012; see reference in Chapter 4), for example, pointed to the potential importance of the Atlantic Multidecadal Oscillation (AMO), suggesting that changes in the circulation and temperature of the North Atlantic, recorded in terms of variations in the related index, could account for much of the interdecadal variability in land surface temperatures over the past 60 years. And there are a variety of additional sources of variability, summarized in Chapter 4, which can contribute to year-to-year, decade-to-decade, or to even longer term fluctuations in climate. To conclude on this basis that "*a substantial fraction of the warming is from natural causes that have nothing to do with mankind*" would appear to require a major leap of faith, a giant step for climate skeptical mankind.

Conclusion

Unsupported by physical argument, there is little merit to this objection. It should be taken for what it is: the opinion of unreconstructed skeptics unburdened by the facts.

WSJ CRITICISM NUMBER 5

"CO_2 *is not a pollutant.* CO_2 *is a colorless and odorless gas, exhaled at high concentrations by each of us, and a key component of the biosphere's life cycle*" (Allegre et al. 2012a). This is partly true but particularly misleading. It is, of course, true that when we take in food and inhale oxygen we exhale CO_2. Oxidation of the carbon in the food we eat—conversion to CO_2—provides the energy source that keeps us alive. The carbon in our food is derived, however, from CO_2 taken up from the atmosphere at most a year or so earlier by photosynthesis (a little longer in the case of meat). When we exhale CO_2, we are simply returning CO_2 to the atmosphere from which it was taken just a short time earlier. The process does not result in any net change, increase or decrease, in the abundance of atmospheric CO_2. In contrast, the carbon in the coal, oil, and natural gas we consume was removed from the atmosphere (in the form of CO_2) several hundred million years ago. When we burn this carbon to extract its energy content, this ancient carbon is returned to the atmosphere. The abundance of CO_2 in the atmosphere increases accordingly. None of this is controversial.

Adding CO_2 to the atmosphere has the indisputable consequence of changing the chemical and radiative properties of the atmosphere with implications for climate. A portion of the added CO_2 is transferred to the ocean, where it is responsible for an increase in acidity. It would be difficult under the circumstances, I would contend, not to consider CO_2 a pollutant. In any event, the question has been resolved definitively as a matter of law. The US Supreme Court ruled that under the terms of the US Clean Air Act, CO_2 and other greenhouses gases may indeed be considered as pollutants. The ruling was important. It means that under powers granted by the Act, the Environmental Protection Agency (EPA) has the authority to regulate emissions of these gases.

Allegre et al. (2012a) argue that "plants do so much better with more CO_2 that greenhouse operators often increase CO_2 concentrations by factors of three or four to get better growth." It is true that under otherwise optimal conditions in a greenhouse, selected plants may do better with more CO_2. But, for plants established in natural, external, environments, their success has less to do with the level of CO_2 to which they are exposed, more with local meteorological conditions (temperature, rainfall, sunlight etc.), the quality of the soil in which they grow, and access to supportive supplies of essential nutrients. The suggestion that higher levels of CO_2 "may have contributed to the great increase in agricultural yields of the past century" (Allegre et al. 2012a) is naïve. It belies the prevailing wisdom that this increase resulted rather from the introduction of high-yielding cereal grains, investments in irrigation systems, and other features of the so-called green revolution. If the suggestion is offered to justify the assertion that CO_2 is not a pollutant, it fails.

Conclusion

There is no merit to the assertion by Allegre et al. (2012a), repeated by Happer (2012), that CO_2 is not a pollutant.

SUMMARY OVERVIEW

I elected in this chapter to focus initially on reactions from nonspecialist friends expressing skepticism as to the significance of the human impact on climate, turning then to the more technical criticisms presented in the op-ed articles published in *The Wall Street Journal* by Allegre et al. (2012a, 2012b) and Happer (2012). I chose not to comment on the more subjective, ad hominem aspects of some of the criticisms advanced in the *WSJ* articles. The suggestion that reputable scientists take public positions on issues motivated primarily by financial self-interest is unworthy, whether the finger is pointed at either proponents or opponents of climate change. I sympathize, though, with the objections raised by Allegre et al. (2012a) with respect to the significance and influence of public statements by professional organizations co-signed by multiple individuals. Complicated scientific issues should be resolved in the conventional fashion, by peer-reviewed papers in reputable scientific journals, not by press releases, not by interviews with sympathetic media, and not by individuals signing on to politically directed statements drafted by others.

I concluded that the technical criticisms voiced by Allegre et al. (2012a) and Happer (2012) were for the most part without merit. These articles, however, have been insidiously influential. They give the impression that there is significant informed scientific disagreement as to whether the threat of human-induced climate change is real. In fact, there is little controversy, at least among those who are professionally competent to have an opinion. As I indicated at the outset, with a few exceptions, the authors of the *WSJ* articles are not climate scientists. The views they express reflect personal convictions rather than insights derived from independent analysis. The criticisms they voice are picked up, however, by talking heads on conservative radio and TV, thus contributing to an unhelpful broad-based distrust of government, science, and intellectual activity more generally. Concern about human-induced climate change is cast as an issue for the liberal fringe. Politicians running for office in the United States, despite views they may hold privately as to the gravity of the issue, carefully avoid bringing the topic up in their public appearances. Too frequently it becomes a subject for ridicule rather than for serious discourse.

There is an urgent need for leaders to lead, to educate, and to have the courage to honestly address complex issues. Courage such as this has been notably lacking in recent years. One can only hope that the political environment will change in the future and that we can have a constructive dialogue on these admittedly complicated issues before it is too late.

Note

1. Columbia University geochemist Wallace Broecker coined this phrase to describe the circulation of waters through the deep regions of the world's oceans. The Conveyor Belt is initiated when cold, saline water sinks from the surface of the North Atlantic in winter. The newly formed deep water flows southward toward Antarctica, where it turns eastward, circling the Antarctic continent, receiving an additional input of dense water from the cold surface ocean surrounding Antarctica. It makes its way then into the Indian and Pacific Oceans. Deep water returns slowly to the surface, with the overall circulation completed by a return flow of surface waters from the Indian and Pacific Oceans to the Atlantic. It takes

approximately a thousand years to complete one loop of this giant Conveyor Belt. There is no reason to expect that the flow of waters in the Conveyor should be steady over this extended time interval. The variations in climate exhibited by the Medieval Climate Optimum and the Little Ice Age could reflect century-scale fluctuations in the properties and function of the Conveyor Belt.

References

Allegre, C., et al. 2012a. No need to panic about global warming. *The Wall Street Journal*. http://online.wsj.com/article/SB10001424052970204301404577171531838421366.html.

Allegre, C., et al. 2012b. Concerned scientists reply on global warming. *The Wall Street Journal*. http://online.wsj.com/article/SB10001424052970203646004577211324408442 9540.html.

Dessler, A. E., and S. C. Sherwood. 2009. Atmospheric science: A matter of humidity. *Science* 323, no. 5917: 1020–1021.

Dessler, A. E., Z. Zhang, et al. 2008. Water-vapor climate feedback inferred from climate fluctuations, 2003–2008. *Geophysical Research Letters* 35, no. 20: 1–5.

Happer, W. 2012. Global warming models are wrong again. *The Wall Street Journal*. http://online.wsj.com/article/SB10001424052702304636404577291352882984274.html.

Trenberth, K., et al. (2012). Check with climate scientists for views on climate. *The Wall Street Journal*. http://online.wsj.com/article/SB10001424052970204740904577193270727472662.html.

6

Coal

ABUNDANT BUT PROBLEMATIC

COAL ACCOUNTED FOR 30.3% of total global energy consumption in 2011, the highest share since 1969 (BP 2012a). Since coal on a per unit energy basis is the most prolific of the fossil fuels in terms of CO_2 emissions, this fact alone underscores the magnitude of the challenge we face in addressing the climate issue. Emissions of CO_2 from oil and natural gas, expressed on a per unit energy basis, amount to approximately 78% and 54% of those from coal. In 2010, 43% of global CO_2 emissions were derived from coal, 36% from oil, and 20% from natural gas.

China was responsible for 49.3% of total coal consumed worldwide in 2011. The United States (13.5%), India (7.9%), and Japan (3.2%) ranked 2 through 4. Consumption in the Asian Pacific region amounted to 71.2% of the global total in 2011, as compared to 14.3% in North America (the United States, Canada, and Mexico) and 13.4% in Europe and Eurasia (including the Russian Federation and Ukraine). Coal accounted for 70% of total energy use in China in 2011 (65.5% in 2013), as compared to 22% in the United States. Use of coal increased in China by 9.7% in 2011 relative to 2010. In contrast, consumption in the United States declined by 0.46% over the same period.

BP (2012b) projects that demand for coal in OECD countries will decrease by 0.8% per year between 2011 and 2030.[1] The projected falloff in OECD countries is offset by growth of 1.9% per year over the same time interval in non-OECD countries. China, in the BP projection, remains the world's largest consumer of coal in 2030 (52% of total global consumption). The growth rate in China is expected to drop, though, from 9% per year over the decade 2000 to 2010, to 3.5% between 2010 and 2020, falling further to 0.4% between 2020 and 2030. The trend, as indicated in the BP analysis, reflects the assumption of a shift to less coal-intensive economic activities, combined with an improvement in overall energy efficiency.

India is projected to surpass the United States in terms of total demand for coal by 2024. The growth rate is anticipated to decrease in this case also, from an annual rate of 6.5% between 2000 and 2010 to an average of 3.6% per year between 2011 and 2030. The slowdown for India is attributed to improvements in energy efficiency offset by rising demand from an expanding industrial sector. Overall global demand for coal between 2011 and 2030 is projected to increase at an average rate of 1.0% per year.

Coal in the United States is used primarily to produce electricity (cf. Fig. 3.1). Applications in developing countries are more diverse, with industrial and domestic consumption accounting for approximately half of total current consumption. BP (2013) projects that the fraction of electric power generated worldwide using coal will decrease from 45% in 2011 to 44% in 2020, dropping further to 39% in 2030, and they envisage a similar, though somewhat smaller, decline in the growth of demand for coal by global industry.

The importance of coal combustion as a source of climate-altering CO_2 and the potential significance of coal-derived sulfate aerosols as an offset to the warming induced by this CO_2 were discussed in Chapter 4. This chapter begins with an account of current patterns of coal use, including a summary of present understanding of the geographic distribution of the resource. It continues with a discussion of what might be identified as the more immediate, negative implications of coal use—the consequences for conventional air pollution, acid rain and emission of mercury. This is followed with a critique of the potential for carbon capture and sequestration, the possibility that we might be able to continue using coal while foregoing emissions to the atmosphere of CO_2. The chapter concludes with a summary of the factors that are likely to influence the future fate of the global coal industry.

NATIONAL RESOURCES AND PATTERNS OF CONSUMPTION

As indicated earlier, China was responsible for close to 50% of total coal consumed globally in 2011. A summary of 2011 consumption data for the top 10 consuming countries together with statistics on production is presented in Table 6.1. Results are presented in units of energy, referenced with respect to a standard defined by the energy contained in a million metric tons of oil equivalent, indicated by the abbreviation mtoe.[2] The table includes a summary of country-specific differences between extraction and consumption, P-C. A positive value for P-C would imply a surplus relative to consumption and would indicate that a country is either committing part of its production to building up reserves in storage, or that it is a net exporter of coal, or a combination of both. Values of P-C are positive for China, as they are also for the United States. There is a difference, however. The United States is actually a net exporter of coal. China is now a net

TABLE 6.1

Consumption and Production of Coal by the 10 Highest Consuming Countries in 2011

Rank	Country	Consumption C (mtoe)	Production P (mtoe)	P-C (mtoe)
1	China	1,839.4	1,956.0	116.6
2	United States	501.9	556.8	54.9
3	India	295.6	222.4	−73.2
4	Japan	117.7	0.7	−117.0
5	South Africa	92.9	143.8	50.9
6	Russian Federation	90.9	157.3	66.4
7	South Korea	79.4	0.9	−78.5
8	Germany	77.6	44.6	−33.0
9	Poland	59.8	56.6	−3.2
10	Australia	49.8	230.8	181.0

Note: Data are presented in units of millions of metric tons of oil equivalent (mtoe) (BP 2012a).

importer. The positive value for P-C for China reflects primarily the role of coal stored domestically in China in anticipation of a future increase in prices for the commodity.[3]

China imported 165 million tons (Mt) of coal in 2010, offset by 19 Mt of exports. A summary of the origins and destinations of imports and exports is presented for China in Figures 6.1 and 6.2. Four countries, Indonesia, Australia, Vietnam, and Mongolia, accounted for 83% of Chinese imports in 2010, broken down as follows: Indonesia, 33%; Australia, 23%; Vietnam, 11%; and Mongolia, 10%. The bulk of Chinese exports went to South Korea (38%) and Japan (34%). The quantity of coal added to supplies in storage amounted to 197.5 Mt.

Countries particularly dependent on imports include Japan and the so-called Asian Tigers, South Korea, Hong Kong, and Taiwan, with imports amounting to 117, 78.5, 47.7, and 41.6 mtoe, respectively, in 2011, accounting for more than 99% of total consumption for each of these countries.[4] Major exporting countries in 2011 included, in rank order, Australia (181.0 mtoe), China (116.6 mtoe), the Russian Federation (66.4 mtoe), United States (54.9 mtoe), Colombia (51.5 mtoe), and South Africa (50.9 mtoe). Exports were responsible for, respectively, 80%, 6%, 42%, 10%, 92%, and 35% of the total quantity of coal produced in these countries in 2011.

As indicated in Table 6.2, the United States ranks number 1 in terms of estimated national reserves of coal. The BP tabulation suggests that as much as 28% of the world's total coal reserves may reside in the United States. The Russian Federation ranks number 2, followed by China, Australia, and India. Table 6.2 includes also entries for the ratio of reserves to production (R/P) for individual countries. These data provide

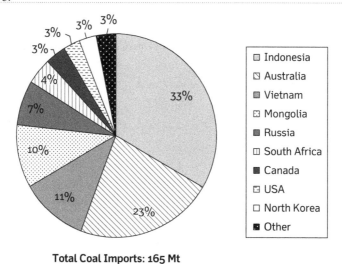

Total Coal Imports: 165 Mt

FIGURE 6.1 China's coal imports in 2010.
(*Source:* LBNL, http://china.lbl.gov/publications/chinaenergy-statistics-2012)

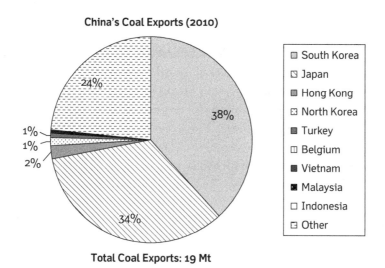

Total Coal Exports: 19 Mt

FIGURE 6.2 China's coal exports in 2010.
(*Source:* LBNL, http://china.lbl.gov/publications/china-energy-statistics-2012)

estimates of how long it would take to deplete a country's estimated reserves, assuming extraction continues at the pace that applied in 2011. A notable feature of the R/P values included in the table is the relatively low value, 33 years, inferred for China. If production of coal in China were to grow at an annual rate of 3.5% as projected by BP (2013) for the 2010–2020 time period, the data in Table 6.2 would suggest that China could run out of domestic supplies of coal by as early as 2032. It is interesting to note in this context, as discussed earlier, that China is now a net importer of coal.

TABLE 6.2

Estimates of Coal Reserves (million tons) and Lifetimes (years) for Available Coal Stocks at Current Depletion Rates for the 10 Best-Endowed Countries (BP 2012a) together with Estimates for the Change in the Abundance of Atmospheric CO_2, Δ (ppm), that would result if total available National Stocks were Consumed

Rank	Country	Anthracite and Bituminous (million tons)	Subbituminous and Lignite (million tons)	Total Reserves (million tons)	R/P (years)	Δ ppm
1	United States	108,501	128,794	237,295	239	32.6
2	Russian Federation	49,088	107,922	157,010	471	20.0
3	China	62,200	52,300	114,500	33	16.4
4	Australia	37,100	39,300	76,400	184	10.6
5	India	56,100	4,500	60,600	103	10.2
6	Germany	99	40,600	40,699	216	4.4
7	Ukraine	15,351	18,522	33,873	390	4.6
8	Kazakhstan	21,500	12,100	33,600	290	5.0
9	South Africa	30,156	—	30,156	118	5.2
10	Colombia	6,366	380	6,746	79	1.1

Table 6.2 includes also an estimate for the increase in the abundance of atmospheric CO_2 that would arise if all of the estimated resources of the 10 largest coal-consuming countries were consumed. The assumption is that 50% of the CO_2 emitted to the atmosphere would persist in the atmosphere: the balance would be incorporated in the ocean consistent with current understanding of the capacity of the ocean to take up excess CO_2 from the atmosphere. According to our assumptions, the 10 largest coal-consuming countries could contribute to an increase of as much as 110 ppm to the abundance of CO_2 in the atmosphere. If BP's (2013) estimate of the total global coal resources were used up over time, the abundance of atmospheric CO_2 could increase by as much as 118 ppm, from approximately 400 ppm today to more than 500 ppm in the future. The projection of 1% per year growth in global coal consumption between 2011 and 2030 (BP 2013) would imply an increase of about 25 ppm by 2030 in the abundance of atmospheric CO_2.

Assumptions:

1. The emission factor for anthracite and bituminous coals is estimated to average 2,690.4 kg CO_2/ton, 1,661.9 kg CO_2/ton for subbituminous and lignite (US EPA, http://www.epa.gov/cpd/pdf/brochure.pdf)
2. 1ppm CO_2 = 7.77 billion tons of CO_2.

COAL AS A SOURCE OF AIR POLLUTION AND ACID RAIN

The environmental problems associated with coal use are both obvious and subtle. Obvious is the impact of the black smoke emitted when the coal consumed is of lower grade. This smoke consists of a complex mixture of soot—black carbon—formed as a product of incomplete combustion, combined with a variety of different mineral compounds. Emissions include also a range of toxic gases, notably sulfur dioxide, nitrogen oxides, and carbon monoxide, and, depending on the origin of the coal, potentially hazardous levels of mercury.

Sulfur and nitrogen oxides, converted to acidic form in the atmosphere, contribute to the phenomenon of acid rain. As the acidified precipitation falls to the surface, it percolates into the ground, triggering decomposition of a variety of naturally occurring minerals in the soil. Included in the elements mobilized in this manner are calcium and magnesium, which are essential for healthy plant growth, and toxic species such as aluminum, which when transported to rivers and lakes can seriously impact the health and even the very survival of fish communities. Problems of fish kills, attributed to acid precipitation, were observed first in rivers and lakes of southern Scandinavia and New England. Acid precipitation has been implicated also in the dieback of forests, especially in regions where the chemical composition of soils is incapable of neutralizing the added acid. Environments underlain by thin soils and granitic bedrock are particularly vulnerable. Soils rich in alkaline minerals such as limestone capable of buffering the acid input are relatively protected.

Sulfur and nitrogen oxides, as discussed earlier, are implicated also in significant production of the reflective, hydroscopic aerosols that contribute to negative radiative forcing (cooling) of the climate system. These particles, typically small in size, have an additional consequential impact on public health. Respired by humans, they can penetrate into the lungs, triggering a variety of serious, potentially life-threatening consequences for impacted communities. While the climatic implications could be perceived as positive, the overall impact of sulfur and nitrogen oxide emissions on exposed natural and human populations is clearly negative. Governments, not only in North America and Europe but also in China, have taken aggressive steps in recent years to limit the scale and scope of these relevant emissions.

Mercury in particulate and oxidized form is removed relatively rapidly from the atmosphere either by incorporation in rain or through contact with surface materials (what atmospheric chemists refer to as dry deposition). In elemental form, though, mercury can survive for a year or more as a gas in the atmosphere. The impact in this case can extend to global scale, in contrast to the more localized impact of the element in its shorter-lived chemical expressions. A significant faction of the mercury emitted by burning mercury-rich coal ends up in aquatic systems. Once there, it can be converted to methyl mercury, the chemical form in which it is most toxic to organisms. Ingested by fish, methyl mercury can be concentrated in the tissue of these organisms. The problem that results is particularly serious for human populations for which mercury-exposed fish constitutes an important fraction of their diet. Methyl mercury is a neurotoxin. Babies born to mothers who have ingested excessive quantities of methyl mercury when pregnant can suffer from a range of disabilities, including difficulties in learning to speak, in processing information, and in coordinating visual and motor functions.

In addition to the range of negative environmental impacts associated with the combustion of coal, it is worth noting the variety of problems associated with its production: the lives that have been lost due to accidents in coalmines; the deaths of coal miners that ensued as a result of black lung and other respiratory diseases; the destruction caused by the removal of entire mountain tops to facilitate more efficient extraction of the resource; the collapse of slag heaps that buried neighboring communities; and the pollution of rivers and ground waters from the disposal of coal ash. Coal may have fueled the success of industrial economies around the world for more than a century, but the benefits have been achieved at significant cost both to human communities and to the environment on which they depend for their health and support.

The obvious problems associated with the use of untreated low-grade coal, the dirty black smoke and the foul-smelling odors, have been recognized for a long time. Freeze (2003) tells the story that when the "bishops and barons and knights from all around England left their country manors and villages and journeyed to London" in the summer of 1306 to participate in the first meeting of the English parliament, they were disgusted and distracted by the "unfamiliar and acrid smell of burning coal." King Edward I responded by banning coal use in the city. Despite this early initiative, coal was established later in

England and elsewhere as the fuel of choice to heat homes and to energize the factories that sprung up in the aftermath of the Industrial Revolution. It took a number of subsequent disasters to persuade policy makers of the need for more aggressive action.

An unusual meteorological situation over London in 1952—an inversion—caused smoke and noxious fumes from the city's homes and factories to accumulate in the atmosphere over a period of 4 days between December 4 and December 8. More than 4,000 people died during this episode, many of them simply falling dead in the streets. Parliament responded by banning the use of soft (smoky) coal in all British cities. A similar incident in Donora, Pennsylvania—in October of 1948—in which up to half of the residents of this small town became sick and a significant number died—had served to raise consciousness earlier in the United States as to the hazards of using untreated coal. Policy responses in the United States included making industrial furnaces more efficient to reduce emission of the carbon compounds responsible for the black soot and the installation of devices—electrostatic precipitators—to remove particulate matter from smoke stacks before it could be released to the air. These steps were eminently successful in addressing the more visible, most obvious aspects of coal-based pollution. If you look at emissions from power plants or other industrial facilities burning coal in the United States or Europe today, the smoke you see is white in color, consisting mainly of water droplets formed from combustion-related water vapor condensing as it cools in leaving the stack. Additional steps must be implemented, however, to cut back on emissions of the sulfur and nitrogen acidic precursors and mercury.

The preferred approach for removing sulfur involves injecting chemicals, calcium oxides for example, into the exhausts of coal-fired furnaces to convert the sulfur to a form that can be recovered prior to emission to the atmosphere (calcium sulfate or gypsum in the case where the injected chemicals involve calcium). Considerable investments of capital are required to provide the necessary equipment. A further problem is that operation of the necessary equipment entails additional expenditure of energy. What that means is that more coal has to be consumed to produce a given amount of electricity.

There are two sources for nitrogen oxides. One stems from the nitrogen present initially in the fuel. The second involves decomposition of atmospheric nitrogen exposed to the high temperatures of the combustion process. It is more difficult to address the latter source in that lowering the temperature of the combustion process will necessarily reduce its overall efficiency. The first can be confronted in principle by reducing the nitrogen content of the fuel prior to combustion. Not surprisingly, measures to reduce emissions of nitrogen oxides from coal-fired industrial facilities have been notably less successful and generally more expensive than those implemented for sulfur. Typically they require installation of expensive catalytic converters capable of selectively removing oxides of nitrogen from the smoke stacks of coal-consuming facilities, similar to the devices now routinely incorporated in the exhaust manifolds of trucks and automobiles in Europe and in the United States.

The US Environmental Protection Agency (EPA) announced ambitious plans in 2012 to limit future emissions of what the Agency classified as hazardous air pollutants (EPA

2012). These include direct emissions of particulate matter, emissions of sulfur and nitrogen oxides and mercury. Details of these requirements can be found at http://www.gpo.gov/fdsys/pkg/FR-2012-02-16/pdf/2012-806.pdf. Technological options are available to meet the stringent standards defined in these regulations. Implementing them will be expensive, however, and will inevitably increase the cost for generation of electricity and other products dependent on coal as the primary source of input energy. By far the most difficult challenge faced by the coal industry, however, involves the climate-mandated requirement to reduce emissions of CO_2.

Is There a Viable Option to Eliminate Emissions of CO_2 from Coal Combustion?

There are three steps that could be implemented to forego emissions to the atmosphere of CO_2 produced by burning coal. First, the CO_2 could be captured from the exhaust stacks of the major industrial facilities burning this coal prior to its emission to the atmosphere. Second, it could be transported to a suitable, assuredly permanent, depository. Finally, it could be buried securely and safely in this depository. The conceptual sequence is referred to as carbon capture and sequestration (CCS).

Carbon dioxide is a relatively minor component of the stack gases emerging from a typical coal-burning power plant or industrial facility. If air is used to burn the coal (the usual practice), molecular nitrogen derived from the incoming air source will represent the most abundant component of the exhaust stream. In addition to N_2, CO_2, and trace amounts of O_2, the exhaust will include significant quantities of particulate matter accompanied by variable quantities of CO, H_2O, SO_2, NO, HCl, HF, mercury, and other metals. The first step in the CCS process requires that the CO_2, present normally at an abundance of less than 14%, must be separated from this complicated chemical exhaust. This can be realized by mixing the exhaust stream with a solvent capable of selective absorption of CO_2. A solution composed of a mixture of water and mono-ethanolamine (C_2H_7NO) has been proposed as a suitable medium to accommodate this requirement. Subsequent steps require reconstitution of the CO_2, which must then be concentrated, purified, and pressurized before it can be transferred to a designated storage reservoir. A variety of options are available potentially to facilitate capture of the carbon produced by combustion of the coal. In all cases, though, they are expensive, not just in terms of demand for capital but also in terms of requirements for the energy needed to capture of this carbon.[5]

Notes

1. The Organization for Economic Co-operation and Development (OECD) represents a grouping of developed countries established in 1961 to promote economic development following free-market principles. Current membership consists of 34 countries. The organization is based in Paris.

2. BP (2012, 2013) elected to express quantities of coal in energy units, selecting the energy content of a million metric tons of oil equivalent (mtoe) as the standard for this purpose: 1 mtoe

corresponds to 39.7 trillion BTU, equal to the energy included in 7.33 million barrels of oil, or 39.2 billion cubic feet of natural gas.

3. Differences between production and consumption inferred from the BP data for any particular year reflect not only changes in exports and imports but also changes in quantities of coal retained in storage.

4. Hong Kong is now officially part of China following transfer of sovereignty from the United Kingdom to China in 1997. The status of Hong Kong is indicated, however, separately in the BP reports, identified as a SAR or Separately Administered Region.

5. The entire process is exceptionally complicated, requiring installation of auxiliary equipment on a scale and cost that could be comparable to the expense involved in constructing the coal-fired facility in the first place. Limitations of space might make it impossible to install and connect the equipment to existing facilities. There is an important energy penalty to be paid also to operate the equipment, which could amount to a minimum of 20% of the energy required to fuel the original facility, most likely as much as a third. Rather than proceeding with expensive retrofits to existing plants to accommodate carbon capture, it might be preferable to incorporate this facility more selectively in new plants, optimizing the design of these plants from the outset to minimize overall costs both for capital and operations. Strategies that could be considered in this case could include the use of concentrated oxygen rather than air as the oxidant for the combustion process, an option referred to as oxyfuel. This would ensure a higher concentration of CO_2 in the exhaust stream since nitrogen would be effectively removed from the oxidant. A second possibility would be to employ the technological option known as integrated gas combined cycle (IGCC) to gasify the coal from the outset before combustion, to convert the energy-bearing carbon of the coal to a gaseous mixture of CO and H_2 with the CO deployed subsequently through reaction with H_2O to yield additional H_2 (through what is known as the water shift reaction). The advantage of the IGCC approach is that it would provide a concentrated stream of CO_2 from the outset but at a significant increase in requirements both for capital and energy. More extensive discussions of the capture process are presented by IPCC (2005) and by McElroy (2010).

References

BP. 2012. *BP statistical review of world energy.* Houston, TX, p. 48.

BP. 2013. *Energy outlook 2030.* Houston, TX, p. 86.

EPA. 2012. National emission standards for hazardous air pollutants from coal- and oil-fired electric utility steam generating units and standards of performance for fossil-fuel-fired electric utility, industrial-commercial-institutional, and small industrial-commercial-institutional steam generating units; Final Rule; 40 CFR Parts 60 and 63, vol. 77, no. 32. https://www.federalregister.gov/articles/2012/02/16/2012-806/national-emission-standards-for-hazardous-air-pollutants-from-coal—and-oil-fired-electric-utility.

IPCC. 2005. *Special report on carbon dioxide capture and storage.* Edited by B. Metz, O. Davidson, H. d. Coninck, M. Loos, and L. Meyer. Geneva, Switzerland: Intergovernmental Panel on Climate Change.

McElroy, M. B. 2010. *Energy perspectives, problems, and prospects.* New York: Oxford University Press.

7

Oil

A VOLATILE PAST, AN UNCERTAIN FUTURE

FOR THOUSANDS OF years, wood was the most important source of energy for human societies. There were many applications for this resource. Arguably, the most important was its role as a source of charcoal, which, burned at a high temperature, made it possible to fashion tools and weapons from copper, tin, bronze, and later iron. When wood ran out, civilizations frequently collapsed, a pattern repeated many times over the course of human history.

Coal replaced wood as the dominant source of energy in England in the early part of the eighteenth century. Benefitting from an advance by Abraham Darby, a Shropshire ironmaster, coal provided the motive force for the Industrial Revolution, which took root at about the same time. Darby's innovation, in 1709, was the development of a protocol to remove impurities such as sulfur from coal that would otherwise have impeded the smelting process. Coke, produced from coal, replaced charcoal, formed from wood, as the critical industrial commodity. Countries rich in coal benefitted accordingly. Only in 1900, however, did coal replace wood as the primary source of energy in the United States, a tribute to the country's abundant sources of timber and the access it enjoyed to a ready source of power available from the series of waterfalls that punctuated the flows of a number of rivers in the country's northeast, notably the Charles River in Massachusetts and the Merrimack River in New Hampshire (including its lower reaches in Massachusetts). As discussed earlier, this latter resource played a pivotal role in the success of the early textile industry in New England.

Oil supplanted coal as the critical global energy source for major industrial economies in the first half of the twentieth century. The roots of oil use extend deep into the past. Oil seeps were exploited in Mesopotamia as early as 5,000 BC to provide a source of asphalt and pitch that was used as mortar to construct the walls and towers of Babylon. Genesis records God's instruction to Noah to "make yourself an ark of gopher wood: make rooms in the ark, and cover it inside and out with pitch." Yergin (1991) quotes the Roman naturalist Pliny in the first century AD extolling the pharmaceutical properties of oil in terms of their ability to stop bleeding, treat maladies as diverse as toothache and diarrhea, and to relieve problems associated with both rheumatism and fever. And the Greeks learned that oil could be processed to develop a fearsome new weapon, a form of napalm known as Greek fire, described by Yergin (1991) as more effective and more terrible than gunpowder. Means to produce and apply this powerful new weapon were guarded as critical state secrets by the Byzantines.

Oil consists of a complex mix of hydrocarbons (molecules composed of a combination of carbon and hydrogen atoms linked to form chains of different length) with variable quantities of sulfur, oxygen, and trace metals. The composition of oil differs from geological reservoir to reservoir. Typically about a third of the hydrocarbons in oil are present in the low mass range, with fewer than 10 carbon atoms per molecule. A third fall in an intermediate range, with between 10 and about 18 carbon atoms per molecule. The heaviest components are represented by viscous, semisolid compounds, as exemplified by tar and asphalt. The key product of the early modern oil industry was kerosene, in wide demand in the pre-electric age as a source of illumination. When combusted, kerosene decomposes into a number of molecular fragments that radiate significant intensities of visible light. The molecules that make up kerosene are composed of between 10 and 15 carbon atoms. Kerosene is slightly heavier and slightly less volatile than gasoline. Gasoline, which together with diesel would emerge as the most important product of the modern oil age, is composed of molecular species consisting of between 8 and 12 carbon atoms: diesel, less volatile than gasoline, is composed of between 10 and 20 carbon atoms.

The birth of the modern oil industry is usually attributed to the successful drilling of a well in Titusville, Pennsylvania, on August 27, 1859. The well struck oil at a depth of 69 feet. The oil had to be pumped to the surface. In short order the key problem was to find a means to store the oil prior to its transport to market. Whiskey barrels provided a convenient temporary solution. To this day, oil is traded in units of barrels: a barrel of oil is defined by a volume of 42 gallons.

John D. Rockefeller was an early entrepreneurial beneficiary of the nascent US oil industry. Together with a partner, Maurice Clark, he built a refinery in Cleveland to convert oil to kerosene, capitalizing on copious supplies of oil available from Titusville and neighboring oil-producing regions of Pennsylvania. Dissolving his partnership with Clark in 1865 for $72,500, he proceeded to build an empire capable of controlling all elements of the oil/kerosene industry, from production, to distillation, to growing timber to make the barrels needed to store the product, to making aggressive deals with railroads to transport his product to east coast markets, to overseeing channels for distribution in the east, even to exploring opportunities for exports to Europe. The Standard Oil Company was formed in 1870 with Rockefeller holding a quarter of the stock. Ten years later, Standard would control more than 90% of total US refining capacity, with dominant positions not only in refining and marketing but also in transportation of the key early product, kerosene. Standard's monopoly ended in 1906 when President Theodore Roosevelt successfully applied the Sherman Antitrust Act of 1890, forcing the company to be broken up into a number of separate units, the largest of which was Standard Oil of New Jersey, forerunner of what is now Exxon Mobil.

By 1910, transportation had replaced lighting as the primary market for oil. The first of the influences driving this transition was Thomas Edison's perfection of a long-lived incandescent light bulb in 1880, combined with subsequent initiatives to ensure a supply of distributed electricity. The second was Henry Ford's success in 1908 in developing and marketing the first affordable automobile, the Model T. Oil continues today to provide the key motive force for global transportation. Each day, motorists in the United States consume more than 400 million gallons of gasoline supplied by processing more than 13 million barrels of oil. Think what would happen were we to run out of oil: cars, trucks, trains, ships, and planes would grind to a halt. And if you didn't live in the countryside—as increasingly most of us do not—you would quickly run out of food. The inevitable result would be panic and widespread starvation.

Following the early developments in Pennsylvania, domestic production of oil grew rapidly in the United States with major discoveries in Ohio, Kansas, California, Texas, Louisiana, and Oklahoma. The strike in Texas in 1901 was so large that the price of oil plunged to as little as 3 cents a barrel, less than the price of a cup of water, according to Yergin (1991). The Texas discovery was surpassed only a few years later by an even greater source from the strike in Oklahoma. International supplies grew apace, with major discoveries in Baku in 1871, Sumatra in 1885, Borneo in 1887, Persia (Iran) in 1908, Mexico in 1910, Venezuela in 1922, Bahrain in 1932, Kuwait in 1938, and, only a few years later, the most significant development of all, recognition of the vast reserves that lay below the desert sands of Saudi Arabia.

The United States was self-sufficient in oil, indeed a major exporter, up to the early 1970s. Subsequently, despite major achievements in exploration, discovery, and exploitation of resources in offshore environments and in Alaska, domestic supplies would prove inadequate to keep up with rapidly increasing domestic demand. After many years of cheap and abundantly available gasoline—25 cents a gallon with rewards for loyalty to specific suppliers when I first arrived in the United States in 1962—complacency finally met reality in 1973. An important contributor to the shock that ensued was the formation of the Organization of the Petroleum Exporters Countries (OPEC) established first in 1960 by Iraq, Kuwait, Iran, Saudi Arabia, and Venezuela, joined later by Libya, United Arab Emirates, Qatar, Indonesia, Algeria, Nigeria, Ecuador, Angola, and Gabon. OPEC would emerge as an important force in setting prices and regulating supplies of oil in the global market, especially in the post-1970s era.

The vulnerability of oil-importing countries to uncertain supplies of foreign oil struck home first in 1973 when Arab members of OPEC, joined by Iran (an OPEC member), Egypt, Syria, and Tunisia, instituted a boycott on exports to selected countries, targeting specifically the United States, Canada, and the Netherlands. The action was intended to punish these countries for their support of Israel during the so-called Yom Kippur War. Hostilities in the war began on October 6, 1970, when Egypt and Syria launched a surprise attack on Israel coincident with Yom Kippur, the holiest day of the Jewish religious calendar. By the time the embargo was lifted 6 months later, the international price of crude had increased from $3 to close to $12 a barrel. A further crisis near the end of the decade was triggered by the overthrow of the Shah in Iran, by the taking of hostages in the US Embassy in Tehran, and by the war that broke out subsequently (in September 1980) between Iraq and Iran, two of the world's largest producers. The Iraq–Iran conflict continued for close to 8 years before a peace treaty was finally agreed on in August 1988. International oil prices in the late 1970s and early 1980s tripled, climbing to a level of close to $34 a barrel, approximately $97 in today's currency, not very different from recent prices, a shock nonetheless to economies long accustomed to low prices and reliable supplies. Historical data on world oil prices are displayed in Figure 7.1, presented both in dollars of the day and 2012-adjusted currency. The data clearly indicate the exceptional volatility that has distinguished global oil markets over the past 40 years.

Successive US administrations, from Nixon to Carter to Reagan to Bush (G.H.W) to Clinton to Bush (G.W.) to Obama, trumpeted plans to promote US energy security. By that they meant encouraging initiatives that could be implemented to reduce or end US

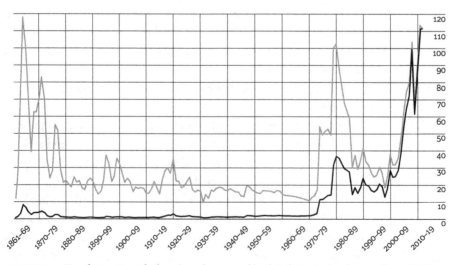

FIGURE 7.1 Trends in prices of oil per barrel expressed both in historical and 2012 adjusted US dollars.
(*Source:* BP 2013)

dependence on imported oil. Fifty years on from the first oil shock, prospects for realization of this shared goal are finally, as we shall discuss, at hand or at least on the horizon for the United States. The outlook for other major oil-importing economies, notably China, is cloudier.

This chapter begins with an overview of the global oil market: which countries are most richly endowed in oil; how rapidly they are depleting their reserves; and given current rates of production, how long it will take before their current reserves are exhausted. It continues with a discussion of recent trends in the oil economy of the United States, highlighting the importance of the increase in domestic production resulting from the recovery of crude oil from shale. The chapter follows with an outline of recent developments in Canada, pointing specifically to the importance of the heavy oil—bitumen—currently held in the extensive tar sands deposits of Alberta. It continues with a discussion of current trends in the oil economy of China, concluding with a summary of key points.

GLOBAL PERSPECTIVES

Table 7.1 presents an account of reserves held by the 10 countries considered by BP (2013) as leading the list of the world's best-endowed oil economies. On a more extensive tabulation, the United States, with reserves estimated at 35 billion barrels, would rank number 11. China, with 17.3 billion barrels, would come in at number 14, trailing Oman (number 13) and Kazakhstan (number 12). The world's total reserve of crude oil is estimated in this survey at 1,668.9 billion barrels. Countries in the Middle East, according to the analysis, account for 48% of the world's total.

The data presented here reflect what BP (2013) refers to as "proved reserves," quantities of oil estimated to be economically recoverable using currently available technology applied to resources that have been previously surveyed. In this sense, the results should be interpreted as at best a lower bound to the quantity of oil that might be available potentially in the future. Highlighting this distinction, we would note the significant difference between estimates of reserves quoted by BP in June 2006 (BP 2006) as compared to those included in the more recent report (BP 2013). The 2006 report rated Venezuela number 6 in terms of specific country reserves for 2005. As indicated in Table 7.1, Venezuela was rated number 1 in the 2013 study (which reported data for calendar year 2012). Canada came in at number 13 in the 2006 survey. It was raised to number 3 in the most recent report. Global reserves for 2005 in the 2006 report were estimated at 1,200.7 billion barrels. This compares with the appraisal for global resources of 1,668.7 billion barrels in the 2013 assessment—an increase of 39%. The differences simply reflect better data on the quantity of potentially exploitable oil held in particular reservoirs and assessment of the advancements in technology available for its extraction.

The change in the estimate of reserves for Canada reflects primarily recognition of the importance of the resource represented by the extensive tar sands deposits of Alberta. The abundance of heavy oil available in the tar sands of the Orinoco basin accounts

TABLE 7.1

Oil Reserves, Ratio of Annual Production to Resources (R/P), and Changes in Concentrations of Atmospheric CO_2 in Parts per Million (Δ ppm) if the Resources Were Completely Consumed, by Country for 2012

Rank	Country	Reserves (billion barrels)	R/P (yr)	Δ ppm
1	Venezuela	297.6	299.1	6.2
2	Saudi Arabia	265.9	63.2	5.5
3	Canada	173.9	127.3	3.6
4	Iran	157.0	116.9	3.2
5	Iraq	150.0	131.9	3.1
6	Kuwait	101.5	88.9	2.1
7	United Arab Emirates	97.8	79.3	2.0
8	Russian Federation	87.2	22.5	1.8
9	Libya	48.0	87.2	1.0
10	Nigeria	37.2	42.2	0.8
	World total	1,668.9	53.1	34.5

Source: BP (2013).

similarly for the upgrade of "proven reserves" reported for Venezuela. Prospects in both cases depend on continuing high prices for oil in the global market.

An ordered list of countries in terms of production in 2012 is presented in Table 7.2. Countries in the Middle East dominate the list, led by Saudi Arabia, which alone accounted for more than 13% of total global production. The Russian Federation, ranked number 8 in terms of reserves, is listed as number 2 in production. The United States and China, the world's largest consumers, placed 3 and 4, respectively. As indicated in Table 7.3, the pattern is clear: the larger the size of an economy, the greater its demand for oil, the sine qua non for success in both developed and developing global economies.

In addition to the data on reserves, Table 7.1 includes an estimate of the time that would elapse before current reserves would be depleted, assuming extraction to continue at its present rate. This is defined by the ratio of reserves to production: R/P expressed in years. Literally interpreted, this would suggest that the world is likely to run out of oil at some point over the next 50 years. The conclusion should be qualified, however, based on the reservations expressed earlier as to the significance of the concept of "proved reserves." If estimates of "proved reserves" should be increased in the future in response to additional data on the resource available in potential deposits, technological advances for their extraction, and higher prices to facilitate this extraction, the conclusions with respect to depletion times should be extended accordingly. As indicated earlier, estimates for "proved reserves" have increased markedly over the past several years, a trend that is likely to continue.

TABLE 7.2

Oil Production by Country for 2012

Rank	Country	Production (million barrels per day)
1	Saudi Arabia	11.53
2	Russian Federation	10.64
3	United States	8.91
4	China	4.16
5	Canada	3.74
6	Iran	3.68
7	United Arab Emirates	3.38
8	Kuwait	3.13
9	Iraq	3.12
10	Mexico	2.91
	World total	86.15

Source: BP (2013).

TABLE 7.3

Oil Consumption by Country for 2012

Rank	Country	Consumption (million barrels per day)
1	United States	18.55
2	China	10.22
3	Japan	4.71
4	India	3.65
5	Russian Federation	3.17
6	Saudi Arabia	2.94
7	Brazil	2.80
8	South Korea	2.46
9	Canada	2.41
10	Germany	2.36
	World total	89.77

Source: BP (2013).

Table 7.1 includes also estimates of the increases in the concentration of atmospheric CO_2 to be expected if all of the currently reported reserves should be consumed in the future. The assumption in this analysis is that 50% of the CO_2 emitted by combustion of oil is likely to remain in the atmosphere for an extended period of time, reflecting what is referred to as the airborne fraction. The balance would be incorporated in the ocean, contributing in this case to an increase in ocean acidity. The data suggest that if current global "proved reserves" were consumed to depletion, the concentration of atmospheric CO_2 would increase by 34.5 ppm from the level of about 400 ppm that prevails today (2013). This would add to the increase of 118 ppm (as discussed in Chapter 6) to be expected if global reserves of coal were also consumed to depletion. Exploitation of the combination of coal and oil could cause the concentration of atmospheric CO_2 to increase to a level in excess of 550 ppm.

US PERSPECTIVES

Important changes are currently underway in the oil economy of the United States. Domestic extraction of crude oil peaked in 1971 at a level of 9.64 million barrels per day (mbd). It fell by 48% between 1971 and 2008 but has since recovered. Production rose by 29.7% between 2008 and 2012, from 5.00 mbd to 6.49 mbd. Imports accounted for 66% of total consumption of crude oil in the United States in 2008. The relative contribution from imports had dropped to 56% in 2012 and has fallen even further subsequently to the point where domestic production of crude oil now exceeds the supply from imports, as indicated in Figure 7.2. The transition from a deficit to a surplus in domestic extraction compared with imports occurred in August of 2013. Two factors contributed to this important reversal of trend: (1) the increase in domestic production indicated in Figure 7.2 and (2) a concurrent decrease in consumption resulting from increases in the economy of fuel use by passenger vehicles dictated by more stringent national Corporate Average Fuel Economy (CAFE) standards.

Extraction of oil from shale was largely responsible for the recent increase in domestic production. The success enjoyed by US industry in this context has been fueled by a

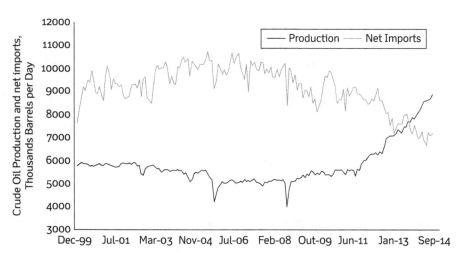

FIGURE 7.2 Monthly domestic production and net imports of crude oil in the United States from January 2000 to September 2014. (http://www.eia.gov/dnav/pet/hist/LeafHandler.ashx?n=PET&s=MCREXUS2&f=A)

combination of high prices for oil and by technological advances allowing for profitable extraction of oil from shale—specifically the ability to drill accurately and reliably not only vertically but also horizontally and to trigger release of hydrocarbons contained in the shale by injection of water and chemicals under high pressure (a process referred to as hydraulic fracturing or fracking, discussed in more detail in the chapter that follows). Key source regions for shale oil in the United States include the Bakken deposit in North Dakota and the Eagle Ford field in South Texas. North Dakota ranked number 2 in terms of individual state production of crude oil in the United States in 2012, surpassed only by Texas. As recent as 2007, it ranked number 8, trailing Texas (number 1), Alaska (number 2), California (number 3), Louisiana (number 4), Oklahoma (number 5), New Mexico (number 6), and Wyoming (number 7). Production from North Dakota increased by 58% between 2011 and 2012—from 0.42 mbd in 2011 to more than 0.66 mbd in 2012. The increase from Eagle Ford was similarly impressive—58% between May 2012 and May 2013. Eagle Ford was responsible for the bulk of the 37% increase in the crude oil production recorded by Texas between 2011 and 2012. Production of oil from shale in the United States is unlikely to decrease significantly in the immediate future.

An equally impressive change is underway in terms of imports and exports of finished petroleum products, as illustrated in Figure 7.3. From a deficit of 0.42 mbd in 2007, the balance for the United States has switched now to a surplus: exports of petroleum products in 2012 exceeded imports by 1.61 mbd. The United States has emerged as the low-cost supplier of many of these products to the global market, taking advantage of the access Gulf Coast refiners enjoy to increasing supplies of heavy crude oil and their ability to process this oil using abundant supplies of inexpensive, domestically produced natural gas (the background to this latter development is discussed in more detail in

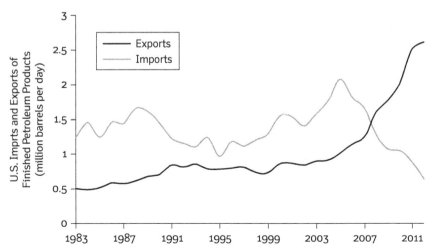

FIGURE 7.3 History of US exports and imports of finished petroleum products. (*Source:* US EIA, 2013a and 2013b)

the chapter that follows). Contributions to US exports of petroleum products in 2012 included the following: distillate fuel oil (diesel), 38.6%; petroleum coke, 19.3%; gasoline, 15.7%; and residual fuel oil, 14.9%, with the balance represented by a suite of products, including notably liquefied petroleum gases. Destination points for these products included, among other countries, Canada, Mexico, Brazil, Chile, the Netherlands, and Venezuela (EIA 2013a and 2013b).

Assuming an average price per barrel of $100, imports of crude oil had a negative impact on US balance of payments of $308.8 billion in 2012. This was offset by a positive net contribution of $71.7 billion from the export of petroleum products, assuming a comparable price per barrel of $100. Approximately 28.6% of the impact of oil trade on the balance of payments may be attributed to the net import of crude oil from Canada. Assuming again a cost per barrel of $100, the net oil trade deficit for the United States in 2012, excluding Canada, would be reduced to $148.6 billion. This may be compared with the deficit for trade with China in 2012 of $436.6 billion.

CANADIAN TAR SANDS

The potential source of recoverable oil in the tar sands of Alberta is estimated at 170 billion barrels, sufficient to accommodate current and anticipated future demand for oil in the United States for at least 30 years. Tar sands account for 99% of the total Alberta oil resource. Production from the tar sands deposits amounted to 1.9 million barrels per day in 2012 and is projected to double by 2022. Tar sands currently account for about half of Canada's total oil production. Approximately 34%, 2.55 million barrels per day, of US net imports of petroleum originated from Canada in 2012. Saudi Arabia and Venezuela were responsible for much of the balance, 18% and 12% respectively, followed by Russia (10%) and Mexico (6%) (Lattanzio 2013).

The hydrocarbon source present in tar sands is not strictly oil but rather a thick, heavy form of bitumen (the technical term for the product) with the consistency of tar (hence the name). There are two strategies available to extract the tar from the sand. Either it can be mined and separated subsequently at the surface, or it can be heated in situ, causing it to flow and allowing it to be piped to the surface. Mining accounted for 51% of production in 2011; 49% was produced using the in situ approach. Both processes are energy intensive. Production of greenhouse gases associated with the in situ approach is generally greater than that involved with the surface mining strategy, reflecting requirements for energy, generally fossil energy, needed to produce the steam and/or electric power employed in extracting the resource in the former case. The in situ strategy is expected to account for an increasing fraction of production in the future.

Two techniques are available to convert the bitumen in the tar sands to a useful form of oil. One involves addition of hydrogen to increase the H to C ratio of the resulting product, usually by reacting the bitumen with a hydrogen-rich compound such as natural gas. The second involves separation of the bitumen into carbon-rich and hydrogen-rich products with subsequent removal of the carbon-rich component (identified under the umbrella designation as coke). Further processing is required to remove sulfur and nitrogen to enhance the quality of the final synthetic crude oil product (SCO). All of these steps require inputs of energy, supplied most commonly in the form of fossil fuels, leading to a significant increase in related emissions of greenhouse gases, notably CO_2.

A major controversy in the United States in recent years relating to the Canadian tar sands resource has concerned the question of whether approval should be given for construction of a pipeline that would provide a direct link between Hardisty, Alberta, and Steele City, Nebraska, thus facilitating delivery of Canadian tar sands products to the United States. Trans Canada, the entity that would construct, own, and operate the proposed pipeline, is a public company domiciled in Canada with diverse energy-related operations—among them, 57,000 kilometers of pipelines dedicated primarily to transportation and distribution of natural gas, storage for a fraction of this gas, and the generation of electric power. It has recently expanded its purvey to include construction and operation of oil pipelines. Keystone, a wholly owned subsidiary, oversees an extensive network of oil-distributing pipelines, including one that already provides an important link between the tar sands of Alberta and markets in the United States.

The existing pipeline extends south from Hardisty, Alberta, proceeds east through Saskatchewan and Manitoba, crosses the border into South Dakota and Nebraska, transitioning east at Steele City, Nebraska, passing through Kansas, Missouri, and Illinois, before ending up in Patoka and Wood River in Illinois. It channels the Albertan oil-sands product to refineries in Illinois; capacity is 590,000 barrels per day. An extension completed in February 2012 delivers a portion of this oil from Steele City to Cushing, Oklahoma, a key distribution center for US crude. A further extension, endorsed by President Obama in March 2012, now under construction, will facilitate transfer of oil from Cushing to refineries on the Gulf of Mexico, reducing the bottleneck for Midwestern crude currently stranded in storage tanks in Cushing. The overall configuration of the existing pipeline network, including the proposed extensions, is depicted in Figure 7.4.

The proposed Keystone XL pipeline would transport 830,000 barrels of oil daily—but not all from the Alberta tar sands: it would also carry at least a portion of the oil derived from the Bakken shale formation in North Dakota. Bakken oil output reached a record of nearly three-quarters of a million barrels per day in April 2013—a 33% increase from the prior year. Approximately 75% of oil produced from the Bakken at present leaves North Dakota by rail. The Keystone extension would only modestly address this imbalance; it would not eliminate the need for additional infrastructure to service the distribution requirements for the rapidly growing production from the Bakken.

Since the proposed pipeline will transit the international border between Canada and the United States, a go-ahead for the project will require an affirmative decision by the US State Department, ultimately by the president. President Obama in a speech on climate policy delivered on June 25, 2013, outlined the ground rules he proposed to

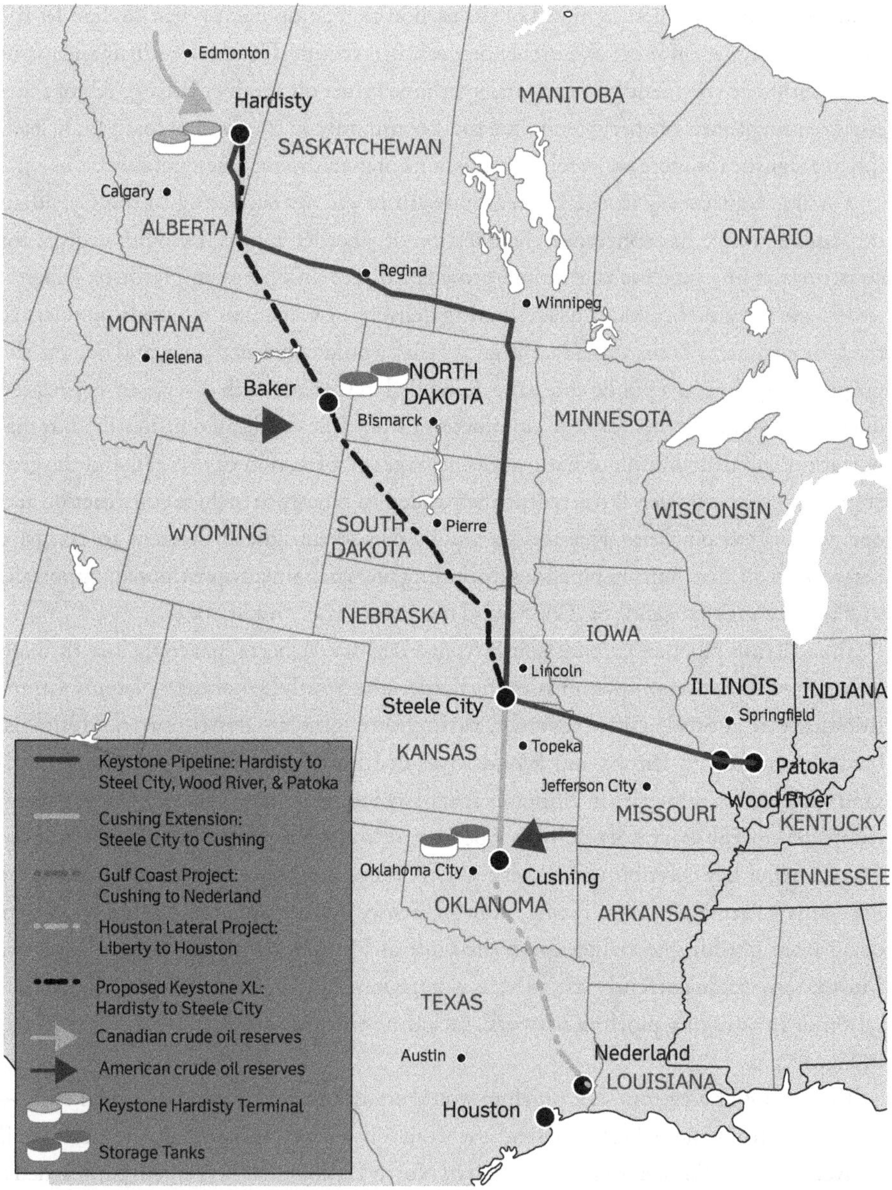

FIGURE 7.4 Existing pipeline network, including proposed extensions.
(*Source:* TransCanada, 2013)

follow in reaching a decision on whether to proceed or not with the project: "allowing the Keystone pipeline to be built requires a finding that doing so would be in the nation's interest. And our national interest will be served only if the project does not significantly exacerbate the problem of carbon pollution." The key question is whether approval of Keystone XL would result in a significant increase in net emissions of greenhouse gases. The critical issue concerns the *net* change in emissions expected to result from the extraction, enhancement, distribution, refining, and ultimate consumption of the tar-sands product as gasoline or diesel fuel in the United States.

Answering this question requires a comprehensive "well-to-wheel" assessment. Lattanzio (2013), in a report prepared recently for the US Congressional Research Service, concluded that per unit of fuel consumed, greenhouse gas emissions (well to wheel) associated with Canadian oil sands would be 14% to 20% higher than a weighted average of transportation fuels now sold or distributed in the United States. He concluded further that "compared to selected imports, Canadian oil-sands crudes range from 9% to 19% more emission-intensive than Middle Eastern Sour, 5% to 13% more emission-intensive than Mexican Maya, and 2% to 18% more emission-intensive than various Venezuelan crudes." Assuming that Keystone XL would deliver a maximum of 830,000 barrels of oil per day to US refineries, he estimated that "incremental pipeline emissions would represent an increase of 0.06% to 0.3% in total annual greenhouse gas emissions for the U.S.," an impact, which though significant, would be scarcely game-changing in terms of the overall implications for human-induced global climate change.

Canada's annual emissions of greenhouse gases grew by 103 million tons between 1990 and 2010, with exploitation of the Alberta tar sands accounting for 32% of this increase. Operations associated with the tar-sands development, including extracting and upgrading the bitumen, accounted for 38.2% of Alberta's CO_2 emissions in 2010. The provincial government has taken steps to improve matters. Under regulations that went into effect in 2010, "large emitters" (operations responsible for annual releases in excess of 50,000 tons of CO_2) must reduce emissions per unit of product by 12% relative to their 2003–2005 baselines. Failure to meet these targets would require delinquent entities to purchase carbon offsets from companies that exceed their quotas or to contribute $15 per ton of excess emissions to a provincial clean-energy fund. (Assets available through this fund totaled more than $300 million Canadian in April 2012.) Emissions could be reduced through increased use of cogeneration to supply the steam and electricity needed to extract the bitumen. Use of renewable resources—solar-powered steam (viable in that location) and wind-derived ancillary electricity, for example—could further reduce greenhouse-gas emissions.

I argued elsewhere (McElroy 2013) that President Obama should approve construction of Keystone XL. His approval, I suggested, could be coupled with a stipulation that the well-to-wheel emissions associated with the tar-sands resource should not exceed the average emissions associated with current use of liquid fuels for transportation in the United States. Increasing supplies of Canadian oil would unquestionably reduce US dependence on potentially unstable and unreliable sources such as Venezuela, Saudi Arabia, and Nigeria. From the US perspective, there are sound economic and security reasons to promote access to Canadian supplies. We are likely to continue for the foreseeable future to require a reliable and affordable supply of oil to fuel our cars, trucks, buses, trains, and planes. With appropriate measures to reduce related emissions of CO_2, damage to the climate system from continued development of the Canadian tar sands resource, while it cannot be eliminated, can at least be minimized. The decision, however, has now been made. Present Obama has elected not to approve the pipeline.

STATUS AND CHALLENGES FOR CHINA

Trends in domestic production and consumption of oil are summarized for China in Figure 7.5. As indicated, China became a net importer of oil in 1993. Despite a modest increase in production since then, averaging approximately 1.9% per year, consumption has outstripped production, growing at an annual average rate of 6.7%. Imports accounted for 60.7% of total consumption in 2012, and the dependence on imports is increasing with time, as illustrated in Figure 7.6. The situation for China may be contrasted with that for the United States, where net imports of oil and oil products accounted for 52.0% of consumption in 2012 and where the current trend, as discussed earlier, favors a secular reduction in the level of net imports.

Patterns for consumption of oil in China differ significantly from those in the United States. In the United States, 71% of oil is used in transportation, 23% by industry, 5% by the residential and commercial sectors, with the balance, 1%, deployed mainly to generate electricity. In contrast, transportation accounted for only 35% of oil consumption in China in 2011, trailing consumption by industry at 39.7%, with residential and commercial uses (9.9%) and agriculture (3.2%) accounting for the bulk of the balance

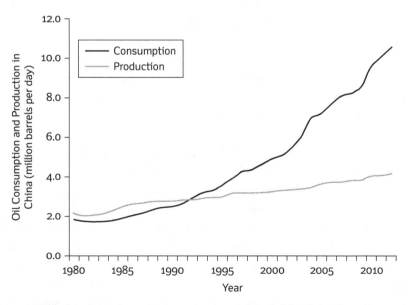

FIGURE 7.5 Trends in domestic production and consumption of oil for China. (http://www.eia.gov/countries/country-data.cfm?fips=CH#pet)

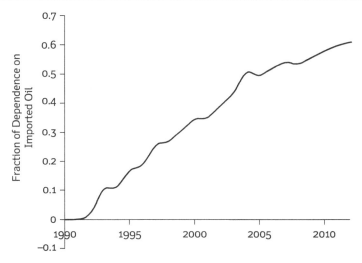

FIGURE 7.6 China's dependence on oil imports. (http://www.eia.gov/countries/country-data.cfm?fips=CH#pet)

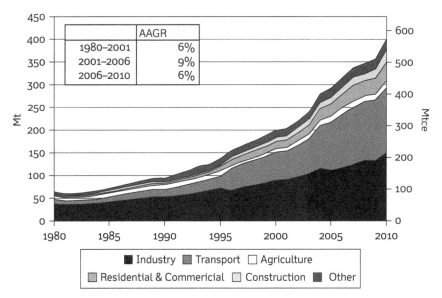

FIGURE 7.7 Oil consumption by sector for China. AAGR refers to the annual average growth rates.
(*Source:* Levine et al. 2013)

(Ma et al. 2012). The contribution from transportation is increasing, however, as indicated in Figure 7.7. There were 74.8 million passenger cars on the road in China in 2011: 17.9 million commercial vehicles. By comparison, in the same year close to 250 million cars and light trucks were registered in the United States.

Middle Eastern sources accounted for 44% of China's oil imports in 2010, led by Saudi Arabia (19%), Iran (9%), Oman (7%), Iraq (5%), and Kuwait (4%). Other important sources included Angola (17%), Russia (6%), and Sudan (5%). In an effort to secure supplies, the major Chinese oil companies, for the most part state owned, have become increasingly active in global oil markets. Taking advantage of China's vast holdings of foreign currency reserves, estimated in excess of $3 trillion in 2012, China's national oil companies (NOCs) have signed oil-for-loan agreements with a variety of countries, including Russia, Kazakhstan, Venezuela, Brazil, Ecuador, Bolivia, Angola, and Ghana. These agreements guarantee supplies of specific amounts oil at specific prices for specific periods of time in return for specific immediate loans. Commitments made under these agreements since 2008 are estimated to have ranged as high as $100 billion (EIA 2013c).

Chinese NOCs have been active also in forming strategic partnerships with international oil companies, in addition to purchasing oil assets in regions as diverse as the Middle East, North America, Latin America, Africa, and Asia (EIA 2013c). The China National Offshore Corporation (CNOOC) signed an agreement in 2012 to purchase the Canadian oil company Nexen. Pending approval from the Canadian government, EIA (2013c) reports that this acquisition will cost at least $15 billion.

The Chinese government is acutely aware of the vulnerability it faces in light of the country's increasing dependence on imports to address its current and future demand for oil. As indicated, it is taking aggressive steps to react to this, by increasing the diversity of supplies, by entering into binding long-term agreements to ensure these supplies, and by investing in both equity and partner relations with potential foreign sources. It is taking steps further to increase the development of domestic resources, both onshore and offshore. Similar to initiatives adopted in other countries, including the United States, China is investing in a strategic reserve, facilities that would store up to 500 million barrels by 2020 to ensure against either disruptive increases in international prices or temporary interruptions in supply. China is acting proactively to address problems that might arise in its future oil economy before they might rise to the level of crisis. It is clearly not only in China's interests but also in the interests of the global community that their efforts should meet with success.

KEY POINTS

1. In its most recent survey, BP rated Venezuela number 1 in terms of proved reserves of oil, supplanting Saudi Arabia. Canada was ranked number 3. The United States and China were rated 11 and 14, respectively.
2. The concept of proved reserves provides at best an imperfect estimate of oil available potentially in the future, conditioned by currently available data on deposits and judgments with respect to technology available for their economically profitable exploitation.
3. The increase in the rankings for Venezuela and Canada reflects primarily recognition of the vast reserves of heavy oil held in the tar sands deposits of

4. There has been an important change in the oil fortunes of the United States over the past several years. Domestic production has increased: the fraction of crude oil derived from imports has decreased from 66% in 2008 to 56% in 2012.
5. Fracking-assisted extraction from shale deposits in North Dakota and South Texas is largely responsible for the US recent trend. Texas and North Dakota now rank 1 and 2, respectively, in terms of state-level domestic production.
6. Availing of cheap prices for domestic natural gas and the ability of refineries on the Gulf Coast to process heavy crude, the United States has emerged now as a net exporter of petroleum products, from a deficit of 420,000 barrels per day in 2007 to a surplus of 1.61 million barrels per day in 2012.
7. China until recently ranked number 2 in terms of both consumption and imports of oil. It has now supplanted the United States as the world's largest importer. In contrast to the situation for the United States, China's fractional dependence on imports is increasing with time, threatening the country's long-term energy security. China is taking aggressive steps, as discussed, to mitigate the challenge posed by this dependence.
8. Increasing supplies of oil, while positive in the short term for global economies, in the longer term pose serious threats to the global climate system.

References

BP. 2006. BP statistical review of world energy. http://www.bp.com/liveassets/bp_internet/russia/bp_russia_english/STAGING/local_assets/downloads_pdfs/s/Stat_Rev_2006_eng.pdf.

BP. 2013. BP statistical review of world energy. http://www.bp.com/en/global/corporate/about-bp/statistical-review-of-world-energy-2013.html.

Lattanzio, R. K. 2013. *Canadian oil sands: Life-cycle assessments of greenhouse gas emissions*. Washington, DC: Congressional Research Service.

Levine, M. D., D. Fridley, et al. 2013. *Key China energy statistics 2012*. Berkeley, CA: Lawrence Berkeley National Laboratory.

Ma, J., W. Zhang, et al., Eds. 2012. *China energy statistical yearbook 2012*. Beijing: China Statistics Press.

McElroy, M. B. 2013. The Keystone XL Pipeline: Should the president approve its construction? *Harvard Magazine* 6: 37–39.

TransCanada. 2013. Keystone XL Pipeline project. http://keystone-xl.com/.

US EIA. 2013a. Exports by destination petroleum and other liquids. Washington, D.C.: US Energy Information Administration: 1. http://www.eia.gov/dnav/pet/pet_move_expc_a_EP00_EEX_mbblpd_a.htm.

US EIA. 2013b. U.S. imports by country of origin petroleum and other liquids. Washington, D.C.: US Energy Information Administration: 1. http://www.eia.gov/dnav/pet/pet_move_impcus_a2_nus_epoo_imo_mbbl_m.htm.

US EIA. 2013c. China. Analysis. Washington, D.C.: US Energy Information Administration: 27. http://www.eia.gov/countries/cab.cfm?fips=CH.

8

Natural Gas

THE LEAST POLLUTING OF THE FOSSIL FUELS

IN TERMS OF emissions from combustion, natural gas, composed mainly of methane (CH_4), is the least polluting of the fossil fuels. Per unit of energy produced, CO_2 emissions from natural gas are 45.7% lower than those from coal (lignite), 27.5% lower than from diesel, and 25.6% lower than from gasoline.

As discussed by Olah et al. (2006), humans have long been aware of the properties of natural gas. Gas leaking out of the ground would frequently catch fire, ignited, for example, by lightning. A leak and a subsequent fire on Mount Parnassus in Greece more than 3,000 years ago prompted the Ancient Greeks to attach mystical properties to the phenomenon—a flame than could burn for a long time without need for an external supply of fuel. They identified the location of this gas leak with the center of the Earth and Universe and built a temple to Apollo to celebrate its unique properties. The temple subsequently became the home for the Oracle of Delphi, celebrated for the prophecies inspired by the temple's perpetual flame.

The first recorded productive use of natural gas was in China, dated at approximately 500 BC. A primitive pipeline constructed using stems of bamboo was deployed to transport gas from its source to a site where it could be used to boil brine to produce both economically valuable salt and potable water. Almost 2,000 years would elapse before natural gas would be tapped for productive use in the West.

Gas from a well drilled near Fredonia, New York, was used to provide an energy source for street lighting in 1821. The Fredonia Gas Light Company, formed in 1858, was the first commercial entity established specifically to market natural gas. Joseph Newton

Pew, founder of the Sun Oil Company (now Sunoco), established a company in 1883 to deliver natural gas to Pittsburgh, where it was used as a substitute for manufactured coal gas (known also as town gas). Pew later sold his interests in natural gas to J. D. Rockefeller's Standard Oil.

The early application of natural gas was primarily for lighting, not only for streets but also for factories and homes. The invention of the Bunsen burner by German chemist Robert Bunsen in 1855 provided the stimulus, near the end of the nineteenth century, for a much wider market, not just for lighting but also for cooking and heating, applications that continue to be of central importance today.

The first long-distance pipeline for distribution of natural gas was established in the United States in 1891. One hundred and twenty miles in length, it delivered natural gas to Chicago from a gas field in central Indiana. By modern standards, this pipeline was primitive. It relied on pressure at the wellhead to propel the gas to its ultimate destination. In contrast, long-distance pipelines today employ compressors to maintain an optimal flow of gas along the transmission lines. The infrastructure for distribution of natural gas in the United States now involves more than 300,000 miles of pipeline with more than 1,400 compressors employed to maintain a reliable, steady flow, with close to 400 underground gas storage facilities deployed to support this distribution network.

The chapter begins with an overview of the international market for natural gas. It provides a list of the 10 countries judged by BP (2013) to host the world's largest proved reserves of natural gas, the largest producers and the largest consumers, together with an illustration of the key distribution channels involved in international trade. It presents estimates for the increase in the concentration of atmospheric CO_2 expected should the estimated reserves of natural gas be consumed to completion, complementing the earlier discussions for coal and oil. It continues with a description of the changing fortunes of the natural gas industry in the United States, highlighting the importance of the increase in extraction from shale made possible through applications of the technology referred to as hydraulic fracturing or fracking. It includes an account of precisely how fracking works and the potential problems associated with its application. The chapter continues with an outline of the current situation for natural gas in China, concluding with a summary of key points of the chapter.

GLOBAL PERSPECTIVES

Table 8.1 presents an accounting of "proved reserves" of natural gas held by the countries ranked by BP (2013) in the top 10 in this category. Results are quoted here in volume metric units: reserves and production/consumption data in units of trillions of cubic meters and billions of cubic meters per year, respectively. Natural gas is traded on the New York Mercantile Exchange (NYMEX) in energy units, in millions of BTU (MMBTU). The

TABLE 8.1

Reserves of Natural Gas, Ratios of Annual Production to Reserves (R/P), and Changes in Concentrations of Atmospheric CO_2, in Parts per Million (ppm), if Resources Were Completely Consumed: for 2012

Rank	Country	Reserves (trillions of cubic meters)	R/P (yr)	Δ ppm
1	Iran	33.6	209.5	4.12
2	Russia	32.9	55.6	4.04
3	Qatar	25.1	159.6	3.07
4	Turkmenistan	17.5	271.9	2.15
5	United States	8.5	12.5	1.04
6	Saudi Arabia	8.2	80.1	1.01
7	United Arab Emirates	6.1	117.9	0.75
8	Venezuela	5.6	169.6	0.68
9	Nigeria	5.2	119.3	0.63
10	Algeria	4.5	55.3	0.55
	World total	187.3	55.7	22.96

Source: BP (2013).

energy content of a million BTU is equivalent to a volume of 28 cubic meters at standard temperature and pressure (STP), equal in turn to a volume of 990 cubic feet. As indicated in Chapter 2, residential bills for natural gas in the United States are quoted in units of therms: 1 therm corresponds to 100,000 BTU.

Iran leads the list in terms of "proved reserves," followed by Russia, Qatar, and Turkmenistan, with the United States in position number 5. Similar to the situation noted earlier for oil, Middle Eastern countries have a dominant position also for natural gas, 43% of the world's total proved reserves according to the data presented here. Ordered lists of countries in terms of production and consumption are presented for calendar year 2012 in Tables 8.2 and 8.3. The United States ranks number 1 in terms of both production and consumption, followed in both cases by Russia.

Middle Eastern countries accounted for 16.3% and 12.4% of total global production and consumption, respectively, in 2012. The difference between production and consumption is reflected in net exports, largely in the form of liquefied natural gas (LNG), with Qatar accounting for the bulk of this trade (78.4%).

Figure 8.1 illustrates current channels for international trade in natural gas, distinguishing between flows by pipeline (gas) as compared to transfers by ship (LNG). The data displayed here indicate that Europe is now critically dependent on imports: 377.2 billion cubic meters in 2012, 34.5% of which originated from Russia, 10.4% from North

TABLE 8.2

Production of Natural Gas by Country for 2012

Rank	Country	Production (billions of cubic meters per year)
1	United States	681.4
2	Russia	592.3
3	Iran	160.5
4	Qatar	157.0
5	Canada	156.5
6	Norway	114.9
7	China	107.2
8	Saudi Arabia	102.8
9	Algeria	81.5
10	Indonesia	71.1
	World total	3,363.9

Source: BP (2013).

TABLE 8.3

Natural Gas Consumption by Country for 2012

Rank	Country	Consumption (billions of cubic meters per year)
1	United States	722.1
2	Russia	416.2
3	Iran	156.1
4	China	143.8
5	Japan	116.7
6	Saudi Arabia	102.8
7	Canada	100.7
8	Mexico	83.7
9	United Kingdom	78.3
10	Germany	75.2
	World total	3,314.4

Source: BP (2013).

Africa (Algeria and Libya). Europe is clearly vulnerable to potential disruption in supplies.

As indicated in Table 8.1, if the volume of natural gas identified by BP (2013) as global "proved reserves" was consumed to depletion, one might expect an increase in the concentration of atmospheric CO_2 by as much as 23 ppm, adding to the increases estimated

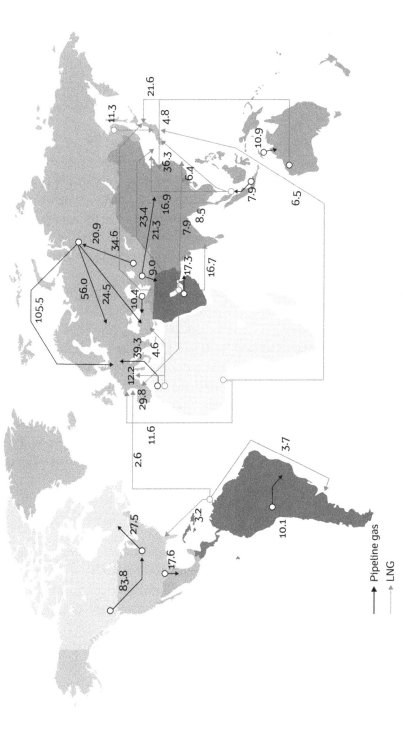

FIGURE 8.1 Paths for international trade in natural gas distinguishing between movements by pipeline (gas) as compared to ship (LNG) with trade values for 2012 expressed in billion cubic meters.
(*Source:* BP Statistical Review of World Energy, June 2013, http://www.bp.com/content/dam/bp/pdf/statistical-review/statistical_review_of_world_energy_2013.pdf)

earlier of 118 ppm from coal and 35 ppm from oil, for a total of 176 ppm. The concentration of atmospheric CO_2 would increase in this case from the present level of about 400 ppm to close to 576 ppm. It should be emphasized, though, as discussed earlier, that the concept of "proved reserves" is at best an imperfect projection of reserves available for recovery in the future (most likely an underestimate). To underscore this point, BP's estimate for global "proved reserves" of natural gas reported for 2005 in their 2006 publication (BP 2006) was revised upward by more than 4% in their most recent analysis for 2012 (BP 2013).

The change (increase) in prospects for natural gas reflected in the different BP reports may be attributed to an important extent to the technology (fracking) deployed over the interim that allows tightly bound gas to be extracted economically from shale. To date, most of this activity has taken place in the United States. It is likely, though, that the technology will be applied extensively to shale deposits elsewhere in the not too distant future. Estimates of "proved reserves" may be expected to rise in this case, requiring upward adjustments in projections for the potential future increase in the concentration of CO_2.

As recently as 10 years ago, it appeared that production of natural gas might have peaked in the United States and that the United States would have to rely on imports to satisfy its continuing, growing demand for this commodity (the excess of imports over domestic extraction peaked between 2004 and 2007). Plans were in place in the late 1990s and early 2000s to construct port facilities that could accommodate an anticipated increase in the supply from imports in the form of LNG. The landscape has changed dramatically over the past several years. The rapid increase in production from shale is largely responsible. Rather than a net importer, the prospect is that the United States could emerge as a net exporter of natural gas in the not too distant future. Port facilities previously designed to accommodate imports of LNG are now being refigured to accommodate exports.

THE SHALE REVOLUTION IN THE UNITED STATES

Shale is a sedimentary rock composed largely of clay minerals (mud) deposited some millions or hundreds of millions of years ago from aquatic environments—the inland seas that covered much of what is now the interior of the United States, for example. As the sediments formed, they incorporated varying amounts of organic matter represented by dead body parts of organisms that once lived in the overlying water column. The higher the biological productivity of the water column, the greater the concentration of dead bodies incorporated ultimately in the underlying sediment.

With increasing deposition of sediment, the surviving organic matter would have been subjected to steadily increasing temperatures and pressures. The original long-chain hydrocarbon molecules that composed the body parts of the dead organisms would have broken down, forming along the way the suite of compounds we associate with oil. Subjected to higher temperatures and pressures (typically at depths in excess of 5 km), only the simplest of these hydrocarbon molecules would survive: methane (CH_4) and low molecular weight compounds such as ethane (C_2H_6), propane (C_3H_8), and pentane (C_5H_{12}). In environments where clay minerals account for a significant fraction of the accumulating sediment, an important fraction of the surviving organic material could end up tightly incorporated in the resulting shale, which forms an effectively impermeable barrier to its potential release. Whether the organic matter preserved in the shale ends up as oil or gas depends on the history of the sedimentary deposit, the range of temperatures and pressures to which it is eventually subjected.

The diversity of shale deposits in the United States is illustrated in Figure 8.2. As indicated in the previous chapter, the resource represented by the Bakken and Eagle Fort deposits is present primarily in the form of oil. Plays dominated by higher molecular weight compounds are classified as wet. Gas-rich deposits are referred to as dry. Gas resources are most abundant in the Marcellus, Haynesville, and Barnett fields, accounting respectively for 55%, 10%, and 6% of estimated recoverable US reserves of gas from shale, with the Barnett-Woodford, Fayetteville, Woodford, Mancos, Eagle Ford, and Antrim plays responsible for the bulk of the balance. Texas led the way in terms of state extraction of natural gas from shale in 2011, followed by Louisiana and Pennsylvania.

Shale gas accounted for 34% of total US gas production in 2011. Its fractional contribution is projected to grow by 50% by 2040, contributing to an increase of 113% in overall domestic production of natural gas (EIA 2013). EIA (2013) projects that the United States could become a net exporter of LNG as early as 2016 and an overall net exporter of natural gas in all forms by 2020.

Multiple steps are involved in the fracking process. The first requires drilling a vertical well to reach the shale, located often as much as several kilometers below the surface: 6,500 to 8,500 feet in the case of the Barnett, even deeper for the Haynesville—10,500 to 13,500 feet. Immediately before reaching the shale, the orientation of the drill bit is shifted from the vertical to the horizontal. Drilling continues horizontally in the shale extending as much as a kilometer or more from the vertical shaft. Following completion of the drilling, the entire well is lined with steel casing, cemented in place with concrete. The horizontal extension of the well is perforated using explosives. Following this, water, sand, and chemicals are injected at high pressure into the well, contacting the shale through the perforations in the casing, triggering a series of small fractures in the rock (hence the name fracking). The sand keeps the cracks open while the chemicals are selected to enhance release of the gas from the shale. A portion of the injected water flows back to the surface as pressure is relieved following completion of the fracking process. The well begins then to produce gas. As many as 25 fracture stages may be involved

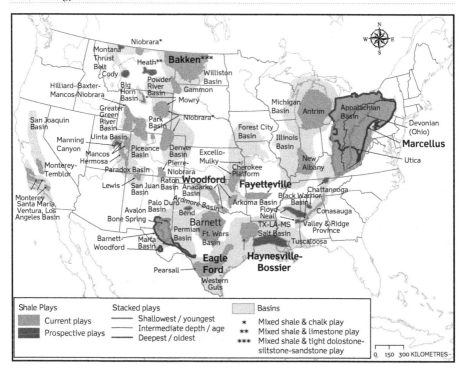

FIGURE 8.2 The locations of primary shale plays in the lower 48 states.
(*Source:* U.S. EIA, http://www.eia.gov/analysis/studies/usshalegas/pdf/usshaleplays.pdf)

in preparing a single site for production, each requiring injection of more than 400,000 gallons of water (a total per well of more than 10 million gallons). A visual rendering of the overall production/fracking operation is presented in Figure 8.3.

George Mitchell, who died on July 20, 2013, at age 94, is generally regarded as the father of the modern fracking industry. Born in Galveston, Texas, the son of poor Greek immigrants (his father was a goat herdsman), Mitchell was a pioneer in the early application of fracking to extract gas from the Barnett shale. He had been successful previously as a gas/oil wildcatter in Texas. He became the first fracking billionaire when he sold his company to the Devon Energy Corporation for $3.5 billion in 2002. Despite his long history in the oil and gas industry, Mitchell was a committed environmentalist and philanthropist. As indicated in his obituary published by *The Guardian* on August 4, 2013, Mitchell built the Woodlands Estate in Texas, widely regarded as a prototype model community; he was a devotee and supporter of astronomy and directed a large portion of his fortune to conservation, including efforts to save the Texas woodpecker.

An award-winning documentary, *Gasland*, produced and directed by Josh Fox in 2010, had an important influence in raising public consciousness as to the potential dangers of fracking. Beginning in Pennsylvania, continuing west to Colorado and

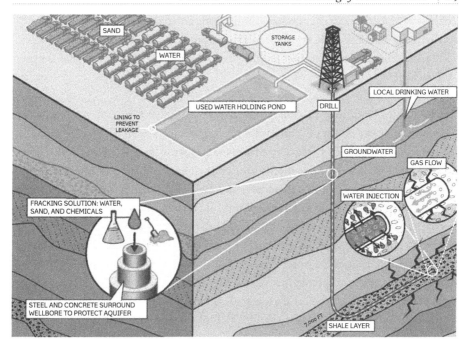

FIGURE 8.3 Illustration of the two phases of hydraulic fracturing: first drilling the wells, then using high-pressure water to fracture the shale and release included gas.
(*Source:* http://harvardmag.com/pdf/2013/01-pdfs/0113-24.pdf)

Wyoming, Fox interviewed a series of individuals claiming to have suffered a variety of serious health and related problems from fracking. A particularly dramatic moment in the documentary records a subject turning on the faucet in his house only to have the water coming out of the tap explode. The water was supplied from the local aquifer. The indication was that the aquifer had been contaminated by chemicals leaking as a result of fracking in the underlying shale.

There are reasons to believe that many of the problems uncovered by *Gasland* reflect irresponsible practices by aggressive operators motivated by opportunities for quick profits with scant regard for longer term responsibilities. Mitchell was a strong believer in the need for government oversight to ensure that fracking should be conducted in an environmentally sensitive fashion. In an interview with *Forbes* reporter Christopher Helman in July 2012 (Helman 2013), he commented: "The administration is trying to tighten up controls. I think it's a good idea. They should have very strict controls. The Department of Energy should do it." He went on to argue:

"There are good techniques to make it (fracking) safe that should be followed properly." But, the smaller, independent drillers "are wild." "It's tough to control these independents. If they do something wrong and dangerous, they should punish them." All of them "know how to set up a proper well and do the proper technology." But a few bad actors could ruin the entire industry. (Quotes from George Mitchell)

If the shale gas industry is to enjoy a long-term profitable future, it is essential that operators gain and retain the confidence of the communities in which they choose to operate. Plans should be in place to ensure that fracking is implemented safely. These plans should include advance monitoring of the chemical composition of the aquifer local residents rely on for their water supplies, continued monitoring during the fracking operation, and an extended commitment to monitoring even after the well has been capped. Credible protocols should be in place to treat fracking fluids once withdrawn from the shale to minimize potential future problems. Steps should be taken to ensure that extraction of gas from shale and its transport to market should not result in an unacceptable increase in fugitive (inadvertent) emissions of methane: as noted earlier, methane on a molecule-per-molecule basis is 10 times more powerful as a climate-altering agent than CO_2. Problems should they emerge should be anticipated, and procedures should be in place in advance to ensure a timely response. All this, of course, will cost money. But as Mitchell pointed out: "any extra costs associated with best practices—assuming all producers follow them—would be passed on in the price of natural gas." The investment is not only necessary but prudent—a down payment essential to ensure the future economic health and viability of the industry.

In the absence of significant channels for either import or export, prices of natural gas have plunged in the United States in response to the increase in production from shale, from a high of $12.69 per million BTU in June 2008 to a record low of $1.82 in April 2012. Trends in international prices for natural gas are displayed for the past 15 years in Figure 8.4 (BP 2013). Note that the spread in prices for different regions was relatively modest up to about 2009. October 2013 prices for landed gas in Japan averaged about $16.00 per million BTU, $15.25 for China, $13.75 for India, and $11.60 for Europe, levels that may be compared with the Henry Hub benchmark price of $3.60 for the United States. The price for natural gas in Canada is similar to that in the United States, reflecting the balancing influence of trade between the two countries.

Cohen (2013), in a comprehensive study of the economics of the shale gas industry in the United States, concluded that the break-even price for production of natural gas from shale is about $4 per million BTU. Producers are benefitting under current market conditions from high prices for associated natural gas liquids (NGLs), notably ethane, propane, and pentane. Prices for NGLs are indexed typically to prevailing international prices for oil. Low prices for natural gas in the United States have been subsidized until

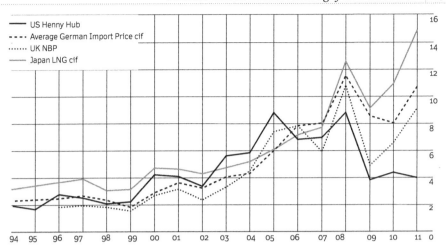

FIGURE 8.4 International prices for natural gas for the past 15 years ($/MMBTU).
(*Source:* BP Statistical Review of World Energy, June 2013, http://www.bp.com/content/dam/bp/pdf/statistical-review/statistical_review_of_world_energy_2013.pdf)

recently by high prices for oil. With continuing high rates for production, prices for NGLs are likely to decline. The subsidy for economically profitable production of natural gas from shale will be reduced accordingly.

If the price for natural gas were the same on an energy basis as the price of oil at $100 a barrel, gas would have to trade at a level of more than $17 per million BTU. Through time, we might expect gas prices to resume their historical relationship with oil. Given the limitations imposed by the current lack of infrastructure to service a major export market for gas in the United States, how might we expect this traditional pattern to be reestablished? The obvious possibilities involve either an increase in domestic consumption, development of the infrastructure required to expand exports, or both. The adjustment is already underway.

Natural gas is increasingly substituting for coal in the US power industry, as indicated in Figure 8.5. Three of the major US railroads, Burlington Northern Santa Fe LLC (BNSF), Union Pacific Corporation, and Norfolk Southern Corporation, have announced plans to turn to natural gas in the form of LNG as a substitute for diesel oil to drive their trains. Warren Buffett indicates that BNSF, a company acquired by Buffet's Berkshire Hathaway Company in 2010, is likely to convert its entire fleet of locomotives to gas in the not too distant future. Major transport companies such as FedEx, UPS, and Waste Management are already modifying their trucks to run on natural gas, this time in the form of compressed natural gas (CNG). And car companies are also following the trend. Ford sold a record number of natural gas–fueled vehicles in 2012 (Mugan 2013). All this is good news, at least in the short term, for US emissions of CO_2. As indicated earlier, CO_2 emissions from combustion of natural gas are 45.7% lower than those from coal.

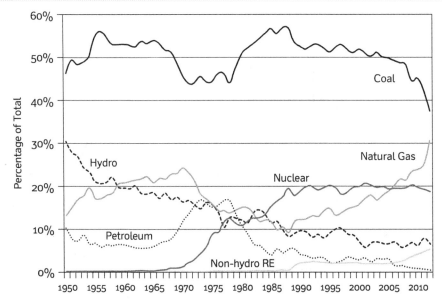

FIGURE 8.5 Increase of natural gas as a substitute for coal in the US power industry. (*Source:* Logan, 2013)

As indicated, the United States is projected to become a net exporter of gas in the form of LNG by 2016 and a net exporter of gas in all forms by 2020, the delay in the latter case reflecting changes in the balance of net pipeline trade with Canada and Mexico (EIA 2013). A $10 billion LNG export plant owned by Cheniere Energy is currently under construction at the company's Sabine Path facility in southwest Louisiana. Exxon Mobil, the largest producer of natural gas in the United States, has proposed to build a second $10 billion plant in the same region. As many as two dozen proposals for construction of LNG export facilities in the United States are currently under consideration by US government authorities. From net importer to prospectively net exporter in less than 30 years, the change in fortunes for the US natural gas industry has been truly remarkable.

PERSPECTIVES ON THE NATURAL GAS ECONOMY OF CHINA

Historical data on production and consumption of natural gas in China are displayed in Figure 8.6. China was self-sufficient in terms of consumption and domestic production prior to 2007. Despite a continuing increase in domestic production, China is now increasingly dependent on imports to balance supply and demand. Approximately half of the imports enter the country in the form of LNG and the balance by pipeline, with an important contribution from the Central Asian Gas Pipeline (CAGP) linking China with the former Soviet Republics of Turkmenistan, Uzbekistan, and Kazakhstan.

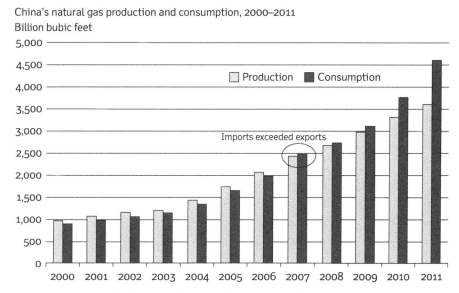

FIGURE 8.6 Historical data on production and consumption of natural gas in China.

Existing and proposed pipelines for distribution of oil and gas in China are illustrated in Figure 8.7.

Natural gas accounts for a relatively small fraction of China's total energy consumption, 5% in 2009 up from 3.4% in 2007, expected to grow, however, to about 10% by 2020. By way of comparison, natural gas was responsible for 26% of total primary energy consumption in the United States in 2011, complemented by petroleum (36%), coal (20%), nuclear (8%), and a combination of hydro, wind, solar, geothermal, and biomass (9%) (EIA 2012). EIA (2013) predicts that Chinese demand for natural gas will more than triple by 2035, increasing at a rate of approximately 5% per year.

Industrial uses accounted for 34% of Chinese consumption of natural gas in 2011. The growth in consumption in recent years reflects primarily demand from the power and residential/commercial sectors, a trend that is likely to continue in the future (EIA 2012). Power generation was responsible for the bulk of natural gas consumption in the United States in 2012 (36%), followed by industry (28%), and a combination of commercial and residential uses (28%), with the balance (8%) associated with transportation and a number of other minor applications.

Domestic production of natural gas in China includes not only the supply from conventional sources—the Sichuan Basin in the southwest, the Tarim, Junggar, and Qaidam Basins in the northwest, and the Ordos Basin in the north—but also a significant, growing contribution from an unconventional source, coal bed methane (CBM). CBM refers to methane incorporated in variable quantities in the solid matrices of coal. China's reserve of CBM is estimated at 35.8 trillion cubic meters, the third largest in

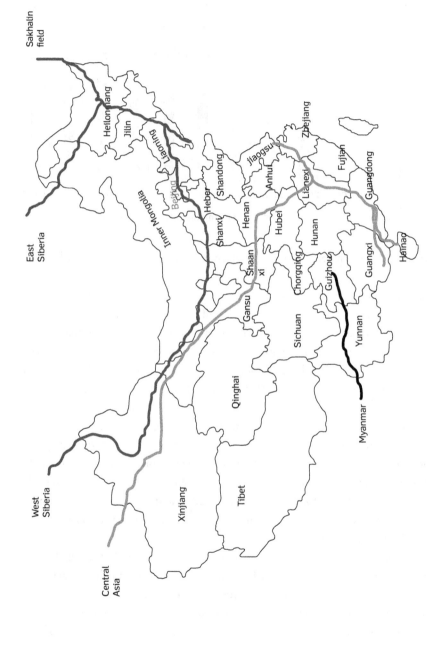

FIGURE 8.7 Existing and proposed pipelines for international supplies of natural gas to China.
(*Source:* PetroMin Pipeliner 2011)

the world, trailing only Russia and Canada, according to a recent report by the official Chinese Xinhua News Agency (http://www.upi.com/Business_News/2013/03/12/China-to-exploit-coal-bed-methane-reserves/UPI-28781363082555/). Production of CBM in China amounted to 315 billion cubic feet per year (Bcf/y) in 2010, accounting for approximately 10% of total domestic gas production. It is projected in the 12th Five-Year Plan to rise to as much as 1,570 Bcf/y by 2030 (EIA 2012). CBM was responsible for 7.3% of dry gas production in the United States in 2011. The relative contribution of CBM to total gas production in the United States is expected to decline in the future, reflecting the increasing importance of the supply from shale.

EIA (2013) estimates that China's reserve of shale gas could be comparable to or even greater than that of the United States. The China National Petroleum Corporation (CNPC) signed a production-sharing contract (PSC) with Royal Dutch Shell in March 2012 to explore development of a block of shale gas deposits in the Sichuan Basin. To date, though, there has been minimal production of gas from shale in China. Despite plans to ramp up production post 2015, pending further more extensive exploration, prospects for future success are at best uncertain.

KEY POINTS

1. Successful application of fracking technology has revolutionized recent production of natural gas in the United States.
2. Prices for natural gas in the United States are three to four times lower than prevailing prices in Europe and Asia.
3. Domestic and industrial consumers in the United States are benefitting significantly from this price differential.
4. EIA (2013) projects that the United States will become a net exporter of natural gas in the form of LNG by 2016 and a net exporter of natural gas in all forms by 2020. Domestic prices are likely to rise accordingly.
5. Continued success of the shale gas industry in the United States will depend on institution of regulatory measures to ensure that production proceeds with minimal damage to local and regional air, soil, and water resources.
6. Steps should be taken to limit inadvertent releases of methane, recognizing the importance of this gas as a climate-altering greenhouse agent.
7. Low prices for natural gas are prompting displacement of coal in favor of gas in the US power sector, contributing to a climate-positive reduction in US emissions of CO_2.
8. Consumption of natural gas is growing rapidly in China. Despite a rise in domestic production, notably in the form of coal bed methane, China is increasingly dependent on imports to meet demand.
9. Shale resources could make an important contribution to China's future demand for natural gas. Pending positive results from ongoing exploratory

activities, future prospects are unclear.

References

BP. 2006. BP statistical review of world energy: 48. http://www.bp.com/liveassets/bp_internet/russia/bp_russia_english/STAGING/local_assets/downloads_pdfs/s/Stat_Rev_2006_eng.pdf.

BP. 2013. BP statistical review of world energy: 48. http://www.bp.com/en/global/corporate/about-bp/statistical-review-of-world-energy-2013.html.

Cohen, A. K. 2013. The shale gas paradox: Assessing the impacts of the shale gas revolution on electricity markets and climate change. Undergraduate Senior Thesis, Harvard University.

Helman, C. 2013. Father of the fracking boom dies—George Mitchell urged greater regulation of drilling. *Forbes*. http://www.forbes.com/sites/christopherhelman/2013/07/27/father-of-the-fracking-boom-dies-george-mitchell-urged-greater-regulation-of-drilling/

Logan, J. 2013. U.S. power sector undergoes dramatic shift in generation mix. Golden, CO: National Renewable Energy Laboratory. https://financere.nrel.gov/finance/content/us-power-sector-undergoes-dramatic-shift-generation-mix#two.

McElroy, M. B., and X. Lu. 2013. Fracking's future: Natural gas, the economy, and America's energy prospects. *Harvard Magazine* January-February: 24–27.

Mugan, S. 2013. Natural gas isn't going to replace gasoline—It already has. *Oil & Energy Daily*. http://www.oilandenergydaily.com/2013/03/27/natural-gas-vehicles-lng-cng/.

Olah, G. A., A. Goeppert, and G. K. S. Prakash. 2006. *Beyond oil and gas: The methanol economy*. Weinheim, Germany: Wiley-VCH Verlag GumbH & Co. KGaA.

PetroMin Pipeliner. 2011. *China's pipeline gas imports: Current situation and outlook to 2025*. Singapore: AP Energy Business Publications.

US EIA. 2013. China. Analysis. Washington, D.C.: U.S. Energy Information Administration. http://www.eia.gov/countries/cab.cfm?fips=CH

9

Nuclear Power

AN OPTIMISTIC BEGINNING, A CLOUDED FUTURE

NUCLEAR POWER WAS widely regarded as the Holy Grail for energy supply when first introduced into the US electricity market in the late 1950s and early 1960s—power so cheap that utilities could scarcely afford the cost of the meters needed to monitor its consumption and charge for its use.[1] The first civilian reactor, with a capacity to produce 60 MW of electricity (MWe), went into service in Shippingport, Pennsylvania, in late 1957.[2] By the end of 1974, 55 reactors were in operation in the United States with a combined capacity of about 32 GWe. The largest individual power plant had a capacity of 1.25 GWe: the capacity of reactors constructed since 1970 averaged more than 1 GWe. The industry then went into a state of suspended animation. A series of highly publicized accidents was responsible for this precipitous change in the fortunes of the industry. Only 13 reactors were ordered in the United States after 1975, and all of these orders were subsequently cancelled.

Public support for nuclear power effectively disappeared in the United States following events that unfolded at the Three Mile Island plant in Pennsylvania on March 28, 1979. It suffered a further setback, not only in the United States but also worldwide, in the wake of the disaster that struck at the Chernobyl nuclear facility in the Ukraine on April 26, 1986. The most recent confidence-sapping development occurred in Japan, at the Fukushima-Daiichi nuclear complex. Floodwaters raised by a tsunami triggered by a major offshore earthquake resulted in a series of self-reinforcing problems in March 2011, culminating in a highly publicized release of radioactivity to the environment that forced the evacuation of more than 300,000 people from the surrounding communities.

If not a death blow, this most recent accident certainly clouded prospects for the future of nuclear power, not only in Japan but also in many other parts of the world. Notably, Germany elected to close down its nuclear facilities, leading to increased dependence on coal to meet its demand for electricity, seriously complicating its objective to markedly reduce the nation's overall emissions of CO_2.

The chapter begins with a primer on nuclear energy—what is it and how is it harnessed to produce electricity. It continues with an account of the developments that led to the accidents at Three Mile Island, Chernobyl, and Fukushim-Daiichi, following with an overview of the current status and future prospects of the nuclear industry, concluding with a summary of the key points of the chapter.

NUCLEAR BASICS

Think of an atom as a microscopic solar system. The bulk of the mass of the atom is contained in the central nucleus (the sun). Think of the electrons that orbit the nucleus as the planets. The analogy is imperfect but nonetheless instructive. The planets in the solar system are bound to the sun by gravity. The negatively charged electrons in the atom are tied to the positively charged nucleus not by gravity but by the electrostatic force responsible for the attraction of particles of opposite electric charge. The number of electrons included in a particular atom depends on the number of positively charged elements, protons, contained in the nucleus. The nucleus of the simplest atom, hydrogen, is composed of a single proton paired with a single electron. More complex atoms include larger numbers of electrons with correspondingly larger numbers of charge-offsetting protons.

The mass of the nucleus of an atom is determined in part by the number of included protons, in part by the number of electrically neutral companion particles known as neutrons. The mass of the neutron is comparable to, though slightly greater than, the mass of the proton. Protons and neutrons are bound together in the nucleus by nuclear forces that are particularly strong and attractive at close separations. A key development in modern physics involved recognition of the basic equivalence of mass and energy. The equivalence is expressed through an equation first derived by Henri Poincaré in 1900 but now more generally attributed to Albert Einstein, who presented a more rigorous formulation of the relationship in 1905. The equation states that energy is equal to mass multiplied by the square of the speed of light: $E = mc^2$. The square of the speed of light is a big number. What this means is that even a small change in mass can result in a large change in energy.

The mass of the nucleus is less than the mass of its component protons and neutrons. The difference reflects the energy with which the protons and neutrons are tied together in the small volume occupied by the nucleus. The system has adjusted to a lower energy state as compared to the conditions that would have applied prior to formation of the nucleus. The decrease in the intrinsic energy of the system is reflected in a corresponding decrease in mass as determined by the Poincaré-Einstein relation. As a consequence, a surplus of energy is released to the external environment.

As discussed briefly in Chapter 2, nuclear reactions in the high-temperature and high-pressure core of the sun are responsible for the energy emitted eventually in the form of visible and ultraviolet light from the outer (colder) regions of the sun's atmosphere. The key initial process here involves fusion, squeezing the protons of hydrogen so close together that they are able to overpower the repulsive (long-range) electrostatic force between these similarly charged particles. This allows the attractive (short-range) nuclear component of the overall interaction to take over, resulting in formation of a new, stable, nuclear complex. Combination of a proton and a neutron in hydrogen leads to the formation of deuterium (D). Subsequent fusion processes involving not only protons but also neutrons are responsible for production of the suite of heavier elements that compose the sun. They are responsible ultimately for the composition of the nebula that formed around the early sun and for the composition of the planets, asteroids, and other bodies that condensed from this nebula.

Despite long-term aspirations for fusion as a potentially inexhaustible and commercially viable source of electricity, the goal remains elusive (see Chapter 2, note 6). Generation of electricity by nuclear processes continues to depend on fission, the release of energy associated with the breakup or fragmentation of unstable nuclei. The uranium atom, key to current applications of nuclear power, is made up of 92 electrons paired with 92 charge-offsetting protons. Uranium is present in natural environments on Earth in three main isotopic forms distinguished by different numbers of neutrons. Individual isotopes are identified by superscripts preceding the appropriate element symbol (U in this case), indicating the sum of the number of protons plus neutrons in their nuclei (the mass number). The most abundant form of uranium is ^{238}U (146 neutrons), accounting for 99.275% of the total supply by number of atoms, followed by ^{235}U (143 neutrons) and ^{234}U (142 neutrons) with relative abundances of 0.72% and 0.0055%, respectively (Bodansky 2004). The fissile component, the isotope that can be induced to break apart (undergo fission) when colliding with a neutron, is ^{235}U.

The fission process is initiated by absorption of a neutron transforming ^{235}U to ^{236}U. Absorption proceeds most readily when ^{235}U is exposed to a low-energy neutron. If the energy of the neutron is high, the neutron is more likely to be captured by the more abundant ^{238}U, resulting in production of plutonium, ^{239}Pu. To enhance the prospects for absorption by ^{235}U, two strategies are employed. First, the supply of uranium to the reactor is enriched in ^{235}U, typically to a level corresponding to a relative abundance of about 3%. Second, reactors include what is referred to as a moderator designed to decrease the energy of the neutrons to the thermal range favoring absorption by ^{235}U. Materials deployed as moderators include normal (light) water, heavy water (water composed mainly of D_2O rather than H_2O), and graphite (carbon). Light water has the advantage that it is readily available and cheap. It suffers from the disadvantage that low-energy neutrons are efficiently captured by H, converted in the process to D. The disadvantage of heavy water is that it is expensive and that it includes inevitably trace concentrations of H. Graphite is more effective than light water as a moderator but less effective than heavy water. The advantage of water, whether light or heavy, is that it can be employed additionally as a coolant. Light water reactors account for 88% of electricity generated currently worldwide by nuclear industry. Systems employing heavy water and graphite make up the balance at levels of 5% and 7%, respectively (Bodansky 2004).

Fuel is supplied to reactors in the form of cylindrical pellets of uranium oxide (UO_2) inserted in tubes constructed from materials selected on the basis of their structural strength and low potential for absorption of neutrons. The tubes are referred to as fuel rods. A typical fuel rod measures 3.7 m in length, 1 cm in diameter. Fuel rods are packaged in units referred to as bundles or assemblies. The Westinghouse pressurized water reactor, a representative example, is loaded with 193 assemblies accommodating 50,952 fuel rods (Bodansky 2004). The assemblies include components, control rods, designed to provide the means to regulate operation of the reactor. The control rods are fabricated from materials, specifically composites of either boron or calcium, selected for their ability to effectively absorb low-energy neutrons. Fully inserted, the control rods have the ability to shut down the reactor.

The ^{236}U isotope produced by absorption of the low-energy neutron is immediately unstable. It breaks apart, forming a spectrum (range) of fresh nuclei accompanied by emission of additional neutrons. A typical reaction sequence could involve, for example, production of an isotope of barium, ^{144}Ba, accompanied by an isotope of krypton, ^{89}Kr. Since the number of protons and neutrons (referred to collectively as nucleons) is conserved (must be the same in the initial and final systems), this reaction is associated with emission of three neutrons, tripling the number of neutrons responsible for the initial fission process. The relevant reaction path may be summarized as follows:

$$n + {}^{235}U \rightarrow {}^{236}U \rightarrow {}^{144}Ba + {}^{89}Kr + 3n \tag{9.1}$$

The combined mass of the products of this reaction is less than the mass of the initial reactants. The difference, according to the Poincaré-Einstein relation, is reflected in a surplus of energy in the products. The bulk of this energy, approximately 96%, is communicated to the product nuclei in the form of kinetic energy with the balance taken up by the accompanying neutrons.[3]

Averaging over all possible reaction paths, fission of ^{235}U results in emission of 2.42, so-called prompt, neutrons. Some of the nuclei formed in the initial fission process are further unstable. They decay on a variety of time scales ranging from 0.2 to 56 sec, averaging about 2 sec, leading to emission of an additional 1.58 (so-called delayed) neutrons per 100 fission events (Bodansky 2004). In a typical light water reactor, for every 10 neutrons absorbed by ^{235}U, approximately 6 neutrons (the higher energy fraction) are absorbed by ^{238}U, leading to production of fissile ^{239}Pu. Capture of an additional neutron results in production of ^{240}Pu, which is subject to further fission. The abundance of ^{239}Pu increases with time as ^{238}U is depleted. Under normal circumstances, fuel assemblies are removed from a reactor and replaced with fresh fuel on a schedule of approximately once every 3 years as the abundance of ^{235}U drops below a critical threshold. Care must be taken to protect the security of the depleted fuel rods after they are removed from the reactor since the ^{239}Pu they include could be used to provide fuel for a nuclear bomb. The bomb dropped on Hiroshima, Japan, on August 6, 1945, was triggered by a high concentration of ^{235}U. ^{239}Pu provided the energy source for the bomb dropped 3 days later on Nagasaki.

Basic elements of the boiling water and pressurized water reactors (BWR and PWR) are illustrated in Figure 9.1(a) and (b). In both cases, the key components—the fuel rods,

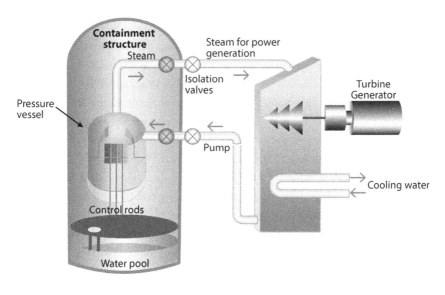

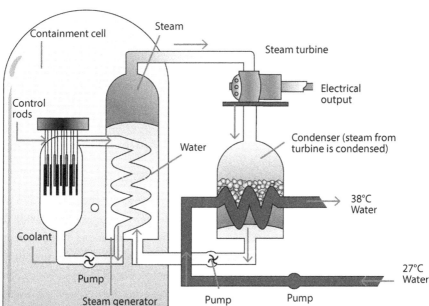

FIGURES 9.1(a) AND (b) Key components of the boiling water and pressurized water reactors (BWR and PWR).
(*Source:* Adapted from Bodansky 2004)

the moderator, and the control rods—are contained in large cylindrical steel tanks, the reactor pressure vessels. The reactor vessels stand approximately 40 feet high with diameters of about 15 feet with walls that in the case of the PWR are capable of withstanding pressures of up to 170 atmospheres (Bodansky 2004). The reactor vessels are enclosed in massive containment buildings that provide important additional levels of security in the event of an accident. Heat from the fission process is converted directly to steam in the BWR, transferred subsequently from the reactor vessel to the turbines used to generate electricity, located, as indicated, outside the containment building. The steam used to generate electricity in the PWR is produced in a separate steam-generating tank placed inside the containment structure but separate from the reactor vessel.

Note in both designs the importance of the steam lines that transfer energy from the fission source to the power-generating turbines and the flows of water that return to the reactor vessel following condensation of the depleted steam. Serious problems would develop should this flow be interrupted. Were the fission process to continue, heat would pile up in the pressure vessels, leading eventually to a breach of the walls of these structures; this would result in emission of a suite of dangerous, highly radioactive, products to the surrounding medium (the containment buildings in the case of the designs illustrated here). In the event of an emergency, the control rods should be inserted to terminate the fission process. The pumps located external to the containment structure should continue to function to provide the water needed to cool the complex.

The spent fuel assemblies once removed from the reactor are immersed in pools of water. They are allowed to cool there before being transferred to steel canisters placed in heavy casks with shielding sufficient to effectively eliminate the risk of any potential release of radiation. The transfer process is handled robotically, safeguarding human operators from any conceivable exposure to radiation. The spent fuel when removed from the reactor is not only hot but also radioactive. The radiation emitted and the heat released decrease by about a factor of 75 over the first year, by a further factor of 5 between years 1 and 10. Reflecting the presence of long-lived radionuclides in the waste stream, emission of radiation remains significant for 100,000 years or even longer (it decreases by a factor of 10,000 over the first 1,000 years, by less than a factor of 100 over the subsequent 100,000 years).

Early plans for nuclear waste in the United States envisaged transfer of spent fuel to a central facility or facilities, where it could be processed for reuse. The product of reprocessing is a mixture of uranium and plutonium oxides referred to as mixed oxide fuel (MOX), which can be used to provide a source of fresh fuel for reactors. Plans for reprocessing in the United States were abandoned in 1977, largely in response to concerns about nuclear proliferation—worries that the plutonium isolated in reprocessing could fall into the wrong hands where it could be used to construct a crude or so-called dirty nuclear bomb. Reprocessing facilities are currently operational in five countries: France, the United Kingdom, Japan, India, and Russia.

While the mass of waste is reduced significantly by reprocessing, reprocessing does little to lower the level of radioactivity. The original fission products are primarily responsible for this radioactivity. There continues to be a need for a depository for long-term storage of nuclear waste. The mass of the material that would have to be delivered to such a depository, however, is significantly less than the mass of the original waste. As much as 98% of the original waste is contributed by uranium, the bulk of which is incorporated in the MOX produced by reprocessing.

After significant study and controversy, Yucca Mountain in Nevada was selected in 1987 as a potential site for long-term storage of nuclear waste in the United States. It was projected that the site would be open to receive waste by 2010. Objections by the Nevada Congressional delegation effectively killed this option. Current plans assume that waste from the US nuclear industry will be stored for the foreseeable future in situ at the plants responsible for its production. Bunn et al. (2001) argued that this strategy was not only cost-effective but also "secure and proliferation resistant."

THREE MILE ISLAND

The accident at Three Mile Island involved one of two pressurized light water reactors (Unit 2, TMI-2) built by Babcock and Wilcox and located on an island in the Susquehanna River approximately 10 miles south of Harrisburg, Pennsylvania. The problems began with a mechanical or electrical failure that cut off the supply of water to the steam generators that provided the primary means for removal of the heat produced by the reactor core. As designed, the turbine-generating system switched off. Control rods were promptly inserted into the fuel assemblies to shut down the reactor. With loss of coolant, pressure began to build up in the reactor. Pressure was relieved by opening a valve, allowing water to flow from the reactor/condenser complex to a tank designed to take up this water. Once the pressure was relieved, the valve was expected to close automatically. It stuck, however, in the open position, permitting water to continue to flow out of the reactor. Due to a faulty design of the systems installed to monitor this situation, the plant operators erroneously assumed that the system had worked as intended. In the absence of an adequate supply of water coolant, the reactor core began to overheat, the materials containing the fuel rods ruptured, and the pellets containing the fuel proceeded to melt. The reactor went into a state of what is referred to as core meltdown. Fortunately, the containment structure did its job restricting for the most part release of radioactive materials to the external environment.

Confusion reigned in the aftermath of the accident in the absence of reliable information on what was actually taking place. The governor of Pennsylvania, Richard Thornburgh, after consulting with officials from the Nuclear Regulatory Commission (NRC), suggested that pregnant women and young children living within 5 miles of the plant should evacuate. As many as 140,000 people took this advice. Three weeks later the majority of these individuals had returned to their homes. President Carter formed a Commission, chaired by John G. Kemeny, president of Dartmouth College, to investigate what had happened and to advise on procedures that could reduce the risk of future problems. The Commission leveled criticisms at essentially all of the parties involved in the Three Mile Island accident, the manufacturer of the plant, the owner, the parties responsible for training the operators, and finally the NRC. Reassuringly, they found that as a result of the accident that "there will either be no case of cancer or the number of cases will be so small that it will never be possible to detect them. The same conclusion applies to other health effects" (Kemeny 1979, 12).

The first human entry to the disabled plant was in July 1980. The top of the reactor vessel was removed in July 1984. Defueling began in October 1985. Off-site shipment of reactor core debris began in July 1986. Defueling was completed in January 1990. Evaporation of contaminated cooling water was initiated in January 1991. Processing of 2.23 million gallons of accident-involved water was finally completed in August 1993, more than 15 years after the accident. Cost for the overall cleanup operations came to more than $1 billion. TMI-2 is now permanently shut down, the radioactive water has been decontaminated, and all of the radioactive debris parts have been shipped to a facility in Idaho operated by the Department of Energy.

CHERNOBYL

The gravity of a nuclear incident or accident is routinely assessed by a group of experts convened under the auspices of the International Atomic Energy Agency (IAEA) and the Nuclear Energy Agency of the Organization for Economic Cooperation on Development (OECD/NEA). Referred to as the International Nuclear and Radiological Event Scale (INES), this process assigns a score of 7 to accidents judged most serious, with scores of 1–3 accorded to events identified as "incidents" rather than "accidents." INES judged the developments at Chernobyl as a 7, a category matched only by what developed later at Fukushima-Daiichi. By way of comparison, the problems at Three Mile Island were assigned a score of 5.

The accident at Chernobyl involved one of four reactors operating on a site located near the city of Pripyat in the Ukraine, then part of the former Soviet Union. It developed in the course of an exercise intended to test the operation of the cooling system for reactor 4. The reactors at the site were equipped with 1,600 fuel rods, cooled individually by throughput of 7,400 gallons of water per hour. The plan was to simulate what would happen if the electricity needed to run the pumps that supplied this water were to fail, requiring the plant to fall back on power provided by on-site auxiliary diesel generators. Power from these generators was expected to be available within 15 seconds. In fact, it took more than 60 seconds before the diesel generators were able to take up the slack. The projection was that power from the plant's steam turbines as they slowed down would be sufficient to make up for this deficit. This did not turn out to be the case. What unfolded was a series of events that included a precipitous increase in power output from the reactor, triggering a steam explosion that sent the 2,000-ton plate to which the entire reactor was fastened through the roof of the reactor building. A second explosion completely destroyed the reactor containment vessel, ejecting radioactive debris to the outside. At the same time the graphite moderator of the reactor ignited, further compounding the problem.

The Soviet authorities were slow to acknowledge the gravity of the events at Chernobyl. The first public recognition that there was a problem came from Sweden, where workers at a Swedish nuclear plant were found to have radioactive particles on their clothes. Radioactivity from Chernobyl had been dispersed by that time widely over eastern and northwestern Europe, triggering widespread, often poorly informed alarms. The city of Pripyat was evacuated quietly 36 hours after the initial explosions, the residents given assurances that they should be able to return within at most a few days. The general population learned of the accident only later through a brief 20-second message on national Soviet TV.

Some 237 individuals, mainly first responders, were diagnosed as suffering from acute radiation sickness in the aftermath of the accident. Over the ensuing months, 28 of these victims would die. The impact of radiation exposure on the larger population appears to have been relatively muted. It has proved difficult to isolate cases of sickness or death that could be attributed unambiguously to radiation exposure over and above what would have been expected to develop under normal, natural circumstances. Follow-up investigations concluded that the design of the Soviet reactors was fatally flawed from the outset. My colleague Richard Wilson, an expert on nuclear power, summed up the situation as follows: these reactors would never have been licensed to operate in the West. The damaged reactor is now enclosed in a massive concrete sarcophagus. The three reactors that continued to operate following the accident have been turned off, shutting down and closing off the entire site. Costs for the cleanup are estimated in the tens of billions of US dollars. There can be little doubt that the damage to public confidence in the safety of nuclear power from the accident at Chernobyl was immediate, profound, and long lasting.

FUKUSHIMA-DAIICHI

The problems at Fukushima-Daiichi began with an earthquake off the east coast of Japan on March 11, 2011. Measuring 8.9 on the Richter scale, this was the fourth largest earthquake recorded worldwide over the past 100 years, the largest ever to strike Japan. It triggered a major tsunami. Floodwaters from the tsunami inundated an extensive region of eastern coastal Japan. As many as 15,700 people lost their lives as a result, 4,650 went missing, 5,300 were injured, 131,000 were misplaced, and 332,400 buildings were destroyed, with serious damage to 2,100 roads, 56 bridges, and 26 railroads (American Nuclear Society 2012).

The Fukushima-Daiichi nuclear complex included six boiling water reactors constructed and installed in the 1970s by the General Electric Company, owned and maintained by the Tokyo Electric Power Company (TEPCO). When the tsunami struck, reactors 5 and 6 were in shutdown mode for scheduled maintenance. Reactor 4 had been defueled. Reactors 1–3 were functioning normally. The complex was protected by a sea wall designed to withstand storm surges as great as 5 meters. Waves from the tsunami crested to heights in excess of 15 meters, causing floodwaters to engulf the entire plant and much of the surrounding region.

Reactors 1–3 shut down as they were designed to in response to the earthquake. The problems that developed later resulted from the loss of the electric power needed to run the pumps required to supply the water used as coolant. Diesel generators that would normally have provided emergency power as backup were flooded and failed to operate. Temperatures built up in the reactors, leading to production of hydrogen gas. A massive explosion in unit 1 triggered by this hydrogen caused the concrete building enclosing the steel reactor vessel to collapse on Saturday, March 12. A further explosion 2 days later blew the roof off the containment structure of unit 3. Unit 4 was seriously damaged in a third explosion that struck on March 15—4 days into the emergency.

There was a limited release of radioactivity in the immediate aftermath of the accident. A more persistent problem involved contamination of ground water leaking through the basements of the plants, interacting with contaminated water that had been used to cool the reactors. As much as 90 million gallons of this water are currently stored in tanks on site, a fraction of which is leaking into the ocean. Operations are currently underway to treat this contaminated water.

In an exceptionally complicated procedure, the fuel rods in unit 4 are now being removed (remember that this material is highly radioactive). Following this, TEPCO plans to extract spent fuel from units 1–3. The entire removal process is expected to take as long as 20–25 years and is anticipated to cost as much as US$100 billion. All but 2 of Japan's original 54 operable nuclear reactors were shuttered (closed down) as of July 1, 2013, placing a heavy burden on conventional fuel sources (coal, oil, and natural gas–fired boilers) to accommodate the country's continuing demand for electric power. Costs for electricity have risen accordingly.

Deficiency in plant design—failure to anticipate the consequences of an earthquake of the magnitude that occurred and the impacts of the ensuing tsunami—played an important role in the series of devastating events that unfolded at Fukushima-Daiichi in 2011. The problems were compounded by the ineptitude of the response by TEPCO and the Japanese authorities. With better information on what was actually happening and with better advance training and deployment of on-site personnel, many of the more serious outcomes from the accident could have been avoided.

Despite the problems and costs, it is important to note that not a single person died as a result of exposure to radiation from the Fukushima-Daiichi accident. The World Health Organization, in a report released in February 2013 (WHO 2013), concluded that, given the level of exposure, the impact in terms of increased incidences of human cancers is likely to be so small as to be undetectable. There can be little doubt, however, that the incident struck yet another blow to the confidence of peoples worldwide as to the safety and security of nuclear power.

CURRENT STATUS

As of July 2013, a total of 427 nuclear reactors were operational in 31 countries worldwide, a decrease from 444 in 2002. Nuclear power was responsible for 17% of electricity generated worldwide in 1993 (the all-time peak). The relative contribution has declined more recently, to a level of 10.4% in 2012 (Schneider and Froggatt 2013). Trends over the past 22 years both in total power generation and in the relative share from nuclear are illustrated in Figure 9.2. An important fraction of the recent decline reflects the changing fortunes for nuclear power in Japan in the wake of the Fukushima-Daiichi disaster. As indicated earlier, following the shutdown of the 10 reactors at Fukushima-Daiichi and the neighboring plant at Fukushima-Daini, only 2 of the remaining 44 reactors in Japan continued to operate after July 1, 2013.

Five countries—in order, the United States, France, Russia, South Korea, and Germany—accounted for 67% of electricity generated worldwide from nuclear sources in 2012. With 100 currently operating reactors, nuclear power was responsible for 19% of electricity generated in the United States in 2012, a decrease from the peak of 22.5% recorded in 1995. In terms of absolute rates for production of electricity from nuclear

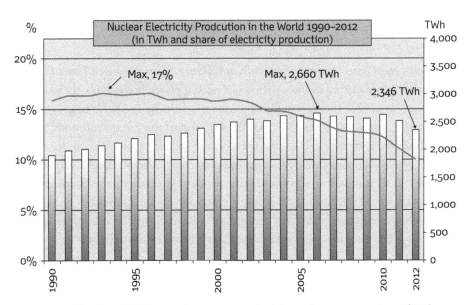

FIGURE 9.2 Trends worldwide over the past 22 years both in total power generation and in the relative share from nuclear.
(*Source:* Schneider and Froggatt 2013)

sources, the United States rated number 1 in 2012, with production almost twice that of France, ranked number 2. France led the world in 2012 in terms of the importance of nuclear power as a fraction of total national power generation (close to 80%), trailed by Slovakia, Belgium, Ukraine, Hungary, Sweden, Slovenia, Switzerland, the Czech Republic, and Finland. On this basis, the United States came in at number 16 and China at number 30 (Schneider and Froggatt 2013).

As of July 2013, there were 18 nuclear reactors operational in China with a combined capacity of 14 GWe with a further 28 reactors under construction. Current plans call for a 5-fold expansion of nuclear capacity by 2020—to between 70 and 80 GWe—with forecasts for additional growth—to 200 GWe by 2030 and to as much as 500 GWe by 2050. To put these numbers in context, it is important to recognize that demand for electricity is expected to more than double in China by 2030 with renewable sources—hydro, wind, and solar—projected to account for more than half of the anticipated expansion. Nuclear sources currently account for about 2% of electric power generation in China. With present plans for expansion, the contribution may be expected to increase to about 11% by 2040 (EIA 2013a).

Despite the ambitious projections for increased supplies of energy from renewable sources and from nuclear, it is likely that China's energy economy will continue to rely primarily on coal for the foreseeable future with an attendant rise in emissions of CO_2. Coal was responsible for 68.4% of total national Chinese primary energy consumption in 2011. Current plans call for the contribution from coal to decrease to about 65% by 2017, a projection assuring continued growth in coal use in the future, albeit at a slower rate than in the past.

FUTURE PROSPECTS

In a recent open letter, climate scientists Ken Caldeira, Kerry Emanuel, James Hansen, and Tom Wigley called for a rapid expansion of nuclear power, in their view the only practical means to combat the threat of disruptive climate change (https://plus.google.com/104173268819779064135/posts/Vs6Csiv1xYr). They argued that "in the real world there is no credible path to climate stabilization that does not include a substantial role for nuclear power." While admitting that "today's nuclear plants are far from perfect," they went on to argue that "passive safety systems and other advances can make new plants much safer. And modern nuclear technology can reduce proliferation risks and solve the waste disposal problem by burning current waste and using fuel more efficiently. Innovation and economies of scale can make new power plants even cheaper than existing plants."

I share their opinion that nuclear power could make an important contribution to the world's future demand for electricity. I doubt though that the transition from fossil fuel–based sources to nuclear can develop as rapidly as they would suggest. Prospects may be better in China, where central planners can make decisions to a large extent unfettered by market forces and where they have the authority to have them implemented. Hard economic realities and public antipathy are likely to limit such progress in the United States and other market-oriented economies, at least in the near term.

There is little evidence to support the view that new nuclear power plants will be "even cheaper than existing plants." EIA (2013b) reports an overnight cost (cost before interest) of $5,429 per kW to build a new nuclear plant in 2012. Depending on interest charges and the time required for construction, the actual cost could be as much as a factor of 2 higher. Moody's credit rating agency describes decisions to invest in large nuclear plants as "bet the firm" challenges for most companies, given that the market capital for even the largest of the publicly traded companies currently operating nuclear plants is less than $50 billion. Overnight capital costs[4] for competitive utility scale power-generating sources are summarized in Table 9.1. The overnight cost for an on-shore wind farm is more than a factor of 2 lower than that for a nuclear plant. Solar thermal and solar photovoltaic plants are also more competitive.

Estimates of levelized costs for different types of electric power-generating sources that could be operational potentially in 2018 are summarized in Table 9.2 (EIA, 2013b). The data included here are intended to reflect national averages. The levelized cost for an on-shore wind farm, for example, is estimated to range from as little as $73.5/MWh for siting at a location where the wind resources are particularly favorable, to as much as $99.8/MWh for siting at a more marginal location. By this measure, overnight or levelized costs for future expansion of nuclear energy would not be competitive.

TABLE 9.1

Overnight Costs for Competitive Utility Scale Power-Generating Sources in 2012 USD

Power Generation Technology	Overnight Capital Cost ($/kW)
Single-unit advanced PC	$3,246
Single-unit advanced PC with CCS	$5,227
Single-unit IGCC	$4,400
Single-unit IGCC with CCS	$6,599
Conventional CC	$917
Conventional CT	$973
Dual-unit nuclear	$5,530
Onshore wind	$2,213
Offshore wind	$6,230
Solar thermal	$5,067
Photovoltaic (150 MW)	$3,873

CC, combined combustion; CCS, carbon capture and storage; CT, combustion turbine; PC, pulverized coal.

Source: EIA (2013, http://www.eia.gov/forecasts/capitalcost/).

TABLE 9.2

Estimates of Levelized Costs for Different Types of Electric Power Sources Operational Potentially in 2019 (2012 US $/1,000 kWh [MWh])
US Average LCOE (2012 $/MWh) for Plants Entering Service in 2019

Plant Type	Capacity Factor (%)	Levelized Capital Cost	Fixed O&M	Variable O&M (including fuel)	Transmission Investment	Total System LCOE	Subsidy[1]	Total LCOE Including Subsidy
DISPATCHABLE TECHNOLOGIES								
Conventional coal	85	60.0	4.2	30.3	1.2	95.6		
Integrated coal-gasification combined cycle (IGCC)	85	76.1	6.9	31.7	1.2	115.9		
IGCC with CCS	85	97.8	9.8	38.6	1.2	147.4		
Natural gas–fired								
Conventional combined cycle	87	14.3	1.7	49.1	1.2	66.3		
Advanced combined cycle	87	15.7	2.0	45.5	1.2	64.4		
Advanced CC with CCS	87	30.3	4.2	55.6	1.2	91.3		
Conventional combustion turbine	30	40.2	2.8	82.0	3.4	128.4		
Advanced combustion turbine	30	27.3	2.7	70.3	3.4	103.8		
Advanced nuclear	90	71.4	11.8	11.8	1.1	96.1	−10.0	86.1
Geothermal	92	34.2	12.2	0.0	1.4	47.9	−3.4	44.5
Biomass	83	47.4	14.5	39.5	1.2	102.6		

Nondispatchable Technologies

Wind	35	64.1	13.0	0.0	3.2	80.3
Wind—offshore	37	175.4	22.8	0.0	5.8	204.1
Solar PV[2]	25	114.5	11.4	0.0	4.1	130.0
Solar thermal	20	195.0	42.1	0.0	6.0	243.1	−11.5	118.6
Hydroelectric[3]	53	72.0	4.1	6.4	2.0	84.5	−19.5	223.6

[1] The subsidy component is based on targeted tax credits such as the production or investment tax credit available for some technologies. It only reflects subsidies available in 2019, which include a permanent 10% investment tax credit for geothermal and solar technologies, and the $18.0/MWh production tax credit for up to 6 GW of advanced nuclear plants, based on the Energy Policy Acts of 1992 and 2005. EIA models tax credit expiration as in current laws and regulations: new solar thermal and PV plants are eligible to receive a 30% investment tax credit on capital expenditures if placed in service before the end of 2016, and 10% thereafter. New wind, geothermal, biomass, hydroelectric, and landfill gas plants are eligible to receive either (1) a $21.5/MWh ($10.7/MWh for technologies other than wind, geothermal, and closed-loop biomass) inflation-adjusted production tax credit over the plant's first 10 years of service or (2) a 30% investment tax credit, if they are under construction before the end of 2013.

[2] Costs are expressed in terms of net AC power available to the grid for the installed capacity.

[3] As modeled, hydroelectric is assumed to have seasonal storage so that it can be dispatched within a season, but overall operation is limited by resources available by site and season.

Source: U.S. Energy information Administration, Annual Energy Outlook 2014 Early Release, December 2013, DOE/EIA-0383ER (2014).

The history of additions to the US electricity-generating system from 1985 to 2011, together with projections of demand for new systems through 2040 (EIA 2013a), is presented in Figure 9.3. As indicated, growth of the US power sector has been dominated over the past 15 years by investments in natural gas–fired systems, complemented more recently by investments in wind. The EIA analysis suggests that the existing network of generating facilities should be sufficient to meet anticipated demand for electricity over the next decade or so. Requirements for new plants are expected to pick up in the mid-2020s in response to the need to replace aging existing facilities. The letter's call for investments in nuclear power as a "credible path to climate stabilization" must be viewed in the context of the data presented in this figure.

If investments in nuclear power were to contribute to a significant reduction in US emissions of CO_2 over the immediate future, it would clearly be necessary to mothball much of the existing gas- and coal-fired plants. While an argument can be made for scrapping many of the older, least efficient, coal-fired facilities, it would be more difficult to justify the elimination of large numbers of the efficient gas-fired systems that came on line more recently. Were such a policy to be implemented, costs for electricity would be expected to skyrocket, with investors, and ultimately consumers, required not only to pay the high price for new nuclear plants but also to absorb the cost for shutting down existing, economically efficient, coal and gas-fired facilities.

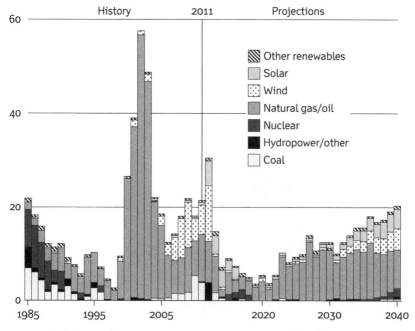

FIGURE 9.3 Additions to electricity generating capacity, 1985–2040 (GW). (*Source:* US EIA, AEO 2013b)

The letter advocated a new generation of nuclear plants that should be proliferation resistant, waste consuming, and passively safe. The goal is laudable, and given time and investment, potentially realizable. It is worth noting though that China's plans for expansion of its nuclear industry are based not on implementation of a radical new approach to plant design but rather on modest improvements to existing technology as indicated by their selection of ACPR1000 and AP1000 as models of choice for the country's nuclear future.

My colleague Joseph Lassiter recommends a renewed commitment by the United States to research and development on what he refers to as New Nuclear, designs that could be established as both cost-effective and safe. He points out that under current regulations, it takes too long and is too expensive (7 to 10 years and more than $500 million) to permit the "rapid experimentation that drives innovation in most of the US economy." He advocates a licensing process modeled after procedures followed by the Federal Drug Administration (FDA) and the Federal Aviation Administration (FAA) in approving new drugs and new aircraft. He suggests that sites should be established at select nuclear reservations in the United States to facilitate experimentation with sales of power from prototypes used to defray expenses. He proposes further "fast track nuclear exports to states with well-established nuclear capabilities and cultures, particularly China and India, negotiating the slowdown and ultimate phasing out of fossil fuel use in their domestic economies as well as their exports to the Rest of the World (ROW)." Professor Lassiter's recommendations merit serious consideration.

Coal-fired and nuclear-fired plants in the present electrical system provide the critical source of what is referred to as baseload power, the steady output needed to meet base-level demand for electricity 24 hours a day, 7 days a week, 365 days a year. Were we to eliminate the source from coal, there would still be a demand for this baseload supply. As discussed later (in Chapter 13), geothermal plants could supply this function in the future. But this option will have to be established on the basis of targeted research and successful experience on operational prototype facilities. In the interim, there would appear to be a continuing essential role for the nuclear option. It is important in this context to note that the nuclear fleet in the United States is aging. Its life can be extended by selective upgrading and by recertification of existing plants. But this, at best, postpones the inevitable, either the need for a new generation of nuclear facilities or an economically viable alternative.

KEY POINTS

1. The energy released by burning coal, oil, or gas involves chemical rearrangement of the atoms in the fuels with the carbon atoms tying up with oxygen to form CO_2, and the hydrogen atoms linking with oxygen to form H_2O. The energy liberated in the nuclear case involves rearrangement of the protons and neutrons that constitute the nuclei of the fuel.

2. Nuclear reactors are fueled by uranium. Uranium is available in geological depositories in three basic isotopic forms. The most abundant is ^{238}U, accounting for 99.275% of supply; followed by ^{235}U, 0.72%; and ^{234}U, 0.0055%. The key component for most civilian reactors is ^{235}U.
3. Fission of uranium leads to production of fast (energetic) neutrons, which are absorbed preferentially by ^{238}U. The energy of the neutrons must be reduced to enhance the likelihood that they should be absorbed by ^{235}U rather than by ^{238}U. Two steps are taken to accomplish this objective. First, the reactors are equipped with what is referred to as a moderator, typically water but in some cases graphite, to decrease the energy of the neutrons. Second, the uranium fuel is treated in advance to enhance (enrich) the relative abundance of ^{235}U.
4. The ^{235}U employed in civilian reactors is enriched typically to about 3%. If ^{235}U is to be deployed as fuel for a nuclear bomb, it must be enriched to a level of 90% or even greater. The current concern with the Iranian nuclear program is that having acquired the expensive centrifuge equipment required to enrich uranium, despite claims to the contrary, the real intent of the Iranian authorities is not simply to produce fuel for civilian reactors but rather to proceed to the high level of enrichment required to manufacture a bomb.
5. Absorption of high-energy neutrons by ^{238}U leads to production of plutonium, which can also be deployed to make a bomb. In a typical reactor, for every 10 neutrons absorbed by ^{235}U, on average 6 are absorbed by ^{238}U, resulting in production of ^{239}Pu. Care must be taken to protect the depleted fuel to ensure the security of this plutonium.
6. The fission process results in the production of a variety of nuclides, many of which are radioactive, producing heat and radiation that can persist for thousands or even hundreds of thousands of years. Current practice in the United States is to store spent fuel on site in the absence of a long-term depository for this waste. In other countries, notably in France, the United Kingdom, Japan, India, and Russia, fuel is reprocessed with a fraction of the reprocessed fuel available for reuse. Reprocessing reduces the volume of waste material. It does not, however, eliminate the need to deal with disposal or management of the long-lived radioactive waste component. The decision not to proceed with reprocessing in the United States was motivated by concern that undesirable parties could subvert the plutonium in the reprocessed fuel and use it to manufacture a bomb.
7. The accidents at Three Mile Island, Chernobyl, and more recently at Fukushima-Daiichi seriously undermined public confidence in the safety of nuclear power. To a large extent, the public unease is based on a lack of understanding of the nature of nuclear power and a lack of appreciation for what actually took place in these accidents. In only one case, Chernobyl, did people

die from exposure to radiation and even in this instance the toll was relatively modest. In contrast, hundreds of thousands, arguably millions, of people have died or suffered serious illness as a result of exposure to air pollution produced by coal-fired alternative sources of electricity.

8. Prospects are bleak for a near-term resurgence of nuclear power in the United States and other OECD countries. The problem relates not just to public unease about safety but also to concerns with expense. Nuclear power is simply not cost-competitive at present with alternative sources of power such as natural gas or even wind. Future growth of the global nuclear industry is most likely to take place in countries such as China, where decisions can be based on factors other than considerations of profit and loss. In China's case, the priority is to ensure a diversity of sources to meet the country's growing demand for electric power, to minimize air pollution problems relating to conventional sources, and to curtail future growth in emissions of CO_2.

9. Nuclear power provides an essential source of baseload power, accounting at present for 19% of total power generation in the United States. The nuclear power plants in the United States are currently aging, pushing the limit on their original design life. The operational life can be extended to some extent by selective upgrading and recertification. But this simply postpones the inevitable demand for an alternative. The need is particularly urgent if we are to successfully transition to a low fossil fuel–based energy future. We need either a new generation of nuclear power plants or an alternative. Geothermal energy could provide an option, but it has not as yet been established.

Notes

1. Lewis L. Strauss, then chairman of the US Atomic Energy Commission, famously remarked in a speech quoted by *The New York Times* on September 17, 1954, that "it is not too much to expect that our children will enjoy electrical energy in their homes too cheap to meter."

2. The symbol *e* included in capacity values is intended to emphasize that the number refers to the capacity for production of electricity. The efficiency for conversion to electricity of the energy released in a typical fission reactor is about 32%.

3. The energy released from fission of a single atom of ^{235}U is greater by a factor of 120 million than the energy released from combustion of a single molecule of CH_4, a dramatic illustration of the dominant importance of nuclear as compared to chemical energy.

4. Overnight capital cost refers to what the plant would cost if built overnight without requirements for financing. Levelized cost provides a convenient measure of the price that would need to be charged for electricity from a specific technology in order for the investment to be profitable. It allows for the cost of building and operating a specific plant over assumed financial and duty cycle lifetimes. The data presented here assume a 30-year cost recovery period with a tax-weighted average cost of capital of 6.6%.

References

American Nuclear Society. 2012. Fukushima-Daiichi: Report by the American Nuclear Society Special Committee on Fukushima, March 2012.

Bodansky, D. 2004. *Nuclear energy: Principles, practices, and prospects*, 2nd ed. New York: Springer.

Bunn, M., J. P. Holdren, A. MacFarlane, S. E. Pickett, A. Suzuki, T. Suzuki, and J. Weeks. 2001. Interim storage of spent nuclear fuel: A safe, flexible, and cost effective near-term approach to spent fuel management. A Joint Report from the Harvard University Project on Managing the Atom and the University of Tokyo Project on Sociotechnics of Nuclear Energy, June 2001.

EIA. 2013a. *Annual energy outlook 2013*. Washington, D.C.: U.S. Energy Information Administration. http://www.eia.gov/forecasts/aeo/pdf/0383(2013).pdf.

EIA. 2013b. *Updated capital cost estimates for utility scale electricity generating plants*. Washington, D.C.: U.S. Energy Information Administration. http://www.eia.gov/forecasts/capitalcost/

Kemeny, J. G. 1979. Report of the President's Commission on the Accident at Three Mile Island: The need for change: The legacy of TMI. October 1979. Washington, D.C.: The Commission.

Schneider, M., and A. Froggatt. 2013. World nuclear industry status report 2013. http://www.worldnuclearreport.org/IMG/pdf/20130716msc-worldnuclearreport2013-lr-v4.pdf.

WHO. 2013. Health risk: Assessment from the nuclear accident after the 2011 Great East Japan Earthquake and Tsunami based on a preliminary dose estimation 2013. http://apps.who.int/iris/bitstream/10665/78218/1/9789241505130_eng.pdf.

10

Power from Wind

OPPORTUNITIES AND CHALLENGES

THE KEY STEP in generating electricity from wind involves capturing and harvesting the kinetic energy of the wind (the energy presented by the directed motion of the air). The blades of a wind turbine are shaped such that the interaction with wind results in a difference in pressure between the top and bottom of the blades. It is this difference in pressure that causes the blades to rotate. And ultimately it is the rotation of the blades that results in the production of electricity.

The physical principle behind the operation of a wind turbine is the same as that that allows a heavy aircraft to stay aloft. The wings of a plane are shaped so that the distance the air has to travel to traverse the underside of the wings is less than the distance it has to move to cross the top. As a result, the flow of air across the top is faster than the flow across the bottom. Bernoulli's Principle states that the greater the speed of the flow, the lower the pressure and vice versa. The difference in pressure between the top and bottom of the wings is what allows the plane to stay aloft (the pressure below is higher, reflecting the lower wind speed). The net upward force exerted by the pressure difference across the wings compensates for the downward pull of gravity, providing the lift that offsets the weight of the plane.

There is a fundamental limit to the extent to which the kinetic energy delivered by the wind can be deployed to turn the blades of the turbine. The absolute limit to the efficiency, derived first by the German physicist Albert Betz and named in his honor (the Betz limit) is 59.3%. With careful design, modern turbines have been able to achieve efficiencies ranging as high as 80% of the Betz limit. They are capable in this case of capturing and making use of as much as 48% of the kinetic energy intercepted by the blades of the turbine and to deploy this power to perform useful functions, most notably to generate electricity.

Wind speeds increase with height above the Earth's surface. To take advantage of this, the rotors for a modern utility-scale turbine—the GE 2.5 MW design, for example—are mounted typically as high as 100 m above the ground (just 7 m lower than the Statue of Liberty, 70 m lower than the Washington Monument) on a tower constructed of conical tubular steel. The blades of the GE 2.5 MW turbine are 50 m long, sweeping out an area of 7,854 m². The power generated by the turbine depends on a number of factors, notably the wind speed and the area swept out by the blades. The kinetic energy contained in a given volume of air is defined by the square of the wind speed. The quantity of kinetic energy crossing a unit area in unit time depends both on the energy content of the wind, proportional to the square of the speed v^2, with an additional factor of v to account for the rate at which this energy is transported through the target area. That is to say, the rate at which energy is delivered to the turbine under ideal conditions depends of the cube of the wind speed, v^3.

The electricity generated as a function of wind speed by the GE 2.5 MW turbine is defined in terms of what is known as the turbine power curve, displayed in Figure 10.1 (GE 2006). The 2.5 MW turbine begins to generate power when the wind speed rises above 3.5 m s^{-1} (7.8 miles per hour). Maximum power output corresponding to the 2.5 MW capacity rating of the turbine is reached at wind speeds greater than about 12.5 m s^{-1}.

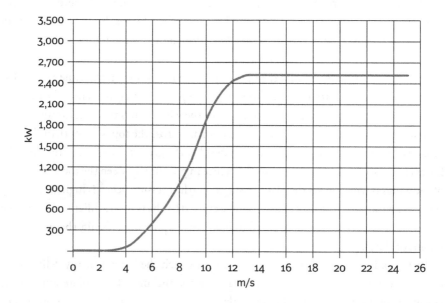

FIGURE 10.1 Power curve for the GE 2.5 MW turbine.
(*Source:* GE 2006)

To avoid damage to the system, the blades of the GE 2.5 MW turbine are feathered at wind speeds higher than 25 m s^{-1} (56 miles per hour).

Turbines in a wind farm are spaced to minimize interference in the airflow between neighboring turbines. This entails a compromise between the objective of maximizing the power generated per turbine and the competing incentive to optimize the number of turbines sited per unit area. Restricting overall power loss to less than 20% requires a downwind spacing of a minimum of 7 rotor diameters with a crosswind spacing of at least 4 rotor diameters (Masters 2004). Applying this criterion to the siting of turbines in a wind farm requires an interturbine areal spacing of one turbine per 0.28 km^2 (between 3 and 4 turbines per square kilometer or 69 acres per turbine).

The chapter begins with an account of the current status of wind power. The perspective is global but with a focus on the United States and China, reflecting the importance of these countries, not only in terms of their contribution to total global energy consumption but also in recognition of the weight of their impact on climate. It continues with an analysis of the potential for wind to meet future global demand for electricity, following with a discussion of problems posed by integrating an intrinsically variable source of power such as wind into a realistic, reliable, electricity supply, concluding with an account of economic incentives that have been implemented to incentivize development of wind resources (and for that matter other carbon-free energy alternatives). It ends with a summary of the key points of the chapter.

CURRENT STATUS

The United States pioneered development of wind-powered generation of electricity in the 1980s and early 1990s. It lost its lead to Europe in the late 1990s as cheap oil, coal, and gas reduced incentives for US utilities to invest in alternative sources of energy. The growth in renewable sources of energy in Europe was fueled to a significant extent by the priority European politicians attached to the issue of climate change, at a time when the US body politic was at best ambivalent. The United States is once again a major player in wind power, joined most recently by China. With 75.6 GW of installed capacity, China ranked number 1 in 2012 in terms of country-level investments in wind. The United States, with 60 GW of capacity, ranked number 2, trailed by Germany (31.3 GW), Spain (22.8GW), and India (18.4 GW) (GWEC 2013).

Table 10.1 provides a summary of the growth in wind capacity for these five countries over the past 16 years. Data on generation of electricity from wind plants installed in the countries are presented in Figure 10.2. Note that for much of the period Germany led the way. The United States supplanted Germany in 2008, with China taking second place a year later. The capacity factors for wind systems in the United States averaged 32.3% in 2013 (http://www.eia.gov/electricity/monthly/epm_table_grapher.cfm?t=epmt_6_07_b) with Germany trailing at about 25%. China's installed capacity in 2012 was approximately 25% greater than that of the United States. Wind accounted for 3.5% of electricity

TABLE 10.1

Historical Data on Wind-Generating Capacity for Top Five Countries (in MW)

	United States	Germany	Spain	China	India
1996	1,590	1,545	249	79	816
1997	1,592	2,080	512	170	950
1998	1,946	2,583	660	224	968
1999	2,500	4,445	1,522	268	1,077
2000	2,566	6,104	2,198	344	1,167
2001	4,261	8,754	3,389	400	1,340
2002	4,685	11,994	4,879	468	1,628
2003	6,374	14,609	6,208	567	1,702
2004	6,740	16,629	8,630	764	2,980
2005	9,149	18,415	10,028	1,260	4,430
2006	11,603	20,622	11,615	2,604	6,270
2007	16,818	22,247	15,145	6,050	8,000
2008	25,170	23,903	16,754	12,210	9,645
2009	35,086	25,777	19,160	25,805	10,925
2010	40,180	27,214	20,676	44733	13065
2011	46,919	29,060	21,674	62,733	16,084
2012	60,007	31,332	22,796	75,564	18,421

Source: http://www.eia.gov/cfapps/ipdbproject/IEDIndex3.cfm?tid=2&pid=2&aid=12.

produced in the United States in 2012, an increase from 0.08% in 2007 (the year in which overall generation of electricity from wind in the United States surpassed production in Germany). Wind was responsible for 1.6% of total power generated in China in 2011. The capacity factor for wind systems in China averaged 21.6% in 2012, rising to 23.7% in 2013. Despite the advantage it enjoys in terms of capacity (Table 10.1), China lags the United States significantly in terms of production of electricity from wind, as indicated in Figure 10.2.

There are two factors that contribute to this aberrant result. The first reflects delays associated with connecting newly installed wind turbines to the grid. The second, more problematic, reflects a circumstance that is arguably unique to China. In China, many residences and commercial establishments are heated in winter with hot water supplied by coal-fired combined heat and power (CHP) plants (facilities designed to deliver a combination of electricity and hot water). These plants are obliged to operate at essentially full capacity in winter in order to meet the demand for hot water. In doing so, they necessarily also produce electricity. This seriously reduces the market for power from wind. As a consequence, wind farms are often idled in China at precisely the time when the output from these facilities should be close to optimal (wind conditions are most favorable in winter). Chen et al. (2014),

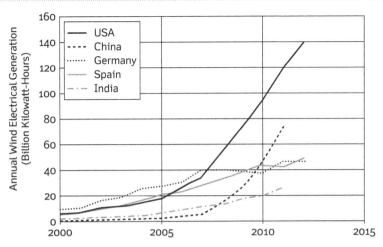

FIGURE 10.2 Historical data on electricity generated from wind in top five countries (in MW) (*Source:* http://www.eia.gov/cfapps/ipdbproject/IEDIndex3.cfm?tid=2&pid=2&aid=12)

in a study of options for meeting the future demand for heat and electricity for Beijing, proposed a potential solution to this dilemma.

Demands for electricity and heat in the Beijing region are projected to increase by 2020 relative to 2009 by 71% and 47%, respectively. If coal were to continue to represent the dominant energy source for both power and heat, emissions of CO_2 would be expected to increase over this interval by 99 million tons (59.6%). Chen et al. (2014) proposed that electricity produced from regional wind resources could provide an important future source for both heat and electricity. Specifically, they argued that if heat for new buildings were supplied by electrically charged heat pumps, the result would be not only an important increase in the utilization of wind but also a significant decrease in emissions of CO_2. If 20% of future expansion in demand for electricity were supplied by wind, savings in CO_2 emissions could amount to as much as 48.5% and could be realized at a relatively modest cost of $15.5 per ton. Even greater savings in emissions could be achieved at higher penetration levels of wind, but at a higher cost: 64.5% reduction in emissions at a penetration level of wind of 40% at a cost of $29.4 per ton.

THE POTENTIAL SOURCE OF POWER FROM WIND

In a study published in 2009, Lu et al. reported estimates for the quantity of electricity that could be generated globally and on a country-by-country basis using wind. Their analysis was based on a record of wind speeds obtained from reanalysis of historical meteorological data,[1] specifically results from Version 5 of the Goddard Earth Observing System Data Assimilation System (GEOS-5 DAS) (Lu et al. 2009). Wind strengths are reported in this reconstruction with a spatial resolution of 2/3° longitude by 1/2° latitude, equivalent to 66.7 km × 50.0 km, with a temporal resolution of 6 hours. Lu et al. (2009) used these data to compute the electricity that could be generated from a broad-scale deployment of 2.5 MW land-based GE turbines. Areas classified as forested, occupied by permanent snow or ice, covered by water, or identified as either developed or urban, were excluded from their analysis. The study considered also power that could be produced using 3.6 MW turbines deployed offshore in waters of depths less than 200 m located within 50 nautical miles (92.6 km) of the nearest shoreline.

Results from the country-by-country analysis are presented in Figure 10.3. Potentials for onshore and offshore production are displayed in Figures 10.3a and 10.3b, respectively (Lu et al. 2009). The results included here were confined to conditions where proposed wind installations could be expected to function with capacity factors (CFs) of no less than 20%[2] (i.e., reasonably good wind conditions). The study suggests that on a global basis a network

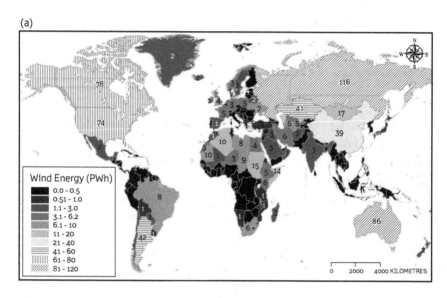

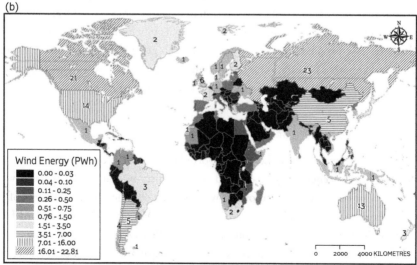

FIGURE 10.3 Annual wind energy potential country by country (A) onshore and (B) offshore.

of 2.5 MW turbines sited on land, with specific regions excluded as discussed, could account for an annual production of 690 PWh (2,353 quad) of electricity.[3] An additional 157 PWh (535 quad) could be obtained from the proposed offshore network of 3.6 MW turbines. To place these results in context, global consumption of electricity amounted to 20.2 PWh (68.9 quad) in 2010. It follows that wind resources available globally, harvested by 2.5 MW land-based turbines operating with CF values higher than 20%, would be more than sufficient to meet current and prospective future global demand for electricity. Important additional production could be available from a network of turbines installed offshore.

A summary of onshore and offshore wind potential for the 10 countries identified as the largest national emitters of CO_2 is presented in Table 10.2.[4] The table includes not only data on emissions of CO_2 but also information on current demand for electricity. In both instances, the data for emissions are quoted for calendar year 2010, and electricity consumption for 2011. The notable feature of the compilation is that, with the exception of Japan, onshore resources of wind are more than sufficient to accommodate present, and most likely projected future, demand for electricity. Offshore resources could make an important contribution to current and future requirements for Japan, compensating for the limited potential for onshore production in this densely populated, tectonically active, mountainous country.

TABLE 10.2

Onshore and Offshore Wind Potential for the 10 Countries Identified as the Largest National Emitters of CO_2

No.	Country	CO_2 Emission (m. tons)	Elec. Consumption (TWh)	Potential Wind Energy (TWh)		
				Onshore	Offshore	Total
1	China	8,547.7	4,207.7	39,000	4,600	44,000
2	United States	5,270.4	3,882.6	74,000	14,000	89,000
3	India	1,830.9	757.9	2,900	1,100	4,000
4	Russia	1,781.7	869.3	120,000	23,000	140,000
5	Japan	1,259.1	983.1	570	2,700	3,200
6	Germany	788.3	537.9	3,200	940	4,100
7	South Korea	657.1	472.2	130	990	1,100
8	Iran	603.6	185.8	5,600	—	5,600
9	Saudi Arabia	582.7	211.6	3,000	—	3,000
10	Canada	499.1	551.6	78,000	21,000	99,000

Note: CO_2 emission for 2012 and electricity consumption for 2011.

Data source: CDIAC (http://cdiac.ornl.gov/trends/emis/meth_reg.html) and US EIA (http://www.eia.gov/beta/international/).

150 | Energy and Climate

The potential for annual wind-generated electricity available from onshore facilities on a state-by-state basis for the United States is summarized in Figure 10.4 (Lu et al. 2009). Note the high concentration of wind resources in the central plains region extending north from Texas to the Dakotas and west from Montana and Wyoming. As indicated in Figure 10.4b, the potential for wind-generated electricity in this region is significantly greater than local demand. In planning for exploitation of this resource, there will be a need for significant upgrading and extension of the existing power transmission grid. It will be important to recognize also that the power potential from wind in this wind-rich region of the country is greatest in winter when demand for electricity is generally

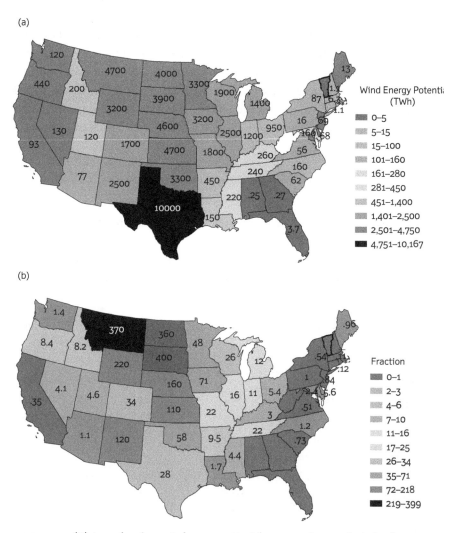

FIGURE 10.4 (A) Annual onshore wind energy potential on a state-by-state basis for the contiguous United States. (B) Same with panel A, but expressed as fraction of total electricity retail sales in the states (2006). For example, the potential source for North Dakota exceeds current total electricity retail sales in that state by a factor of 360.

at a minimum. The seasonal mismatch between demand for electricity and the potential supply from wind is illustrated on a national basis for the contiguous United States in Figure 10.5.

There are problems with and objections to the development of any and all forms of commercial-scale electric power generation systems. Few people would willingly volunteer to have a major coal-fired plant or a gas or nuclear powered plant located in their neighborhood. Not surprisingly, the wind option has also drawn critics. Objections in this case have focused on a combination of concerns about noise and aesthetics. There are two sources of noise, one mechanical, originating from operation of the gearbox, hydraulics, and generator components of the systems; the second, a typical "swooshing" sound, is produced as the air flows around the blades and the tower. The first has been largely eliminated in modern systems by soundproofing the relevant components. The second has been reduced by improvements in blade design, in the process increasing the efficiency with which incident wind energy is converted to power-generating rotary motion of the turbine (http://www.noblepower.com/faqs/documents/06-08-23NEP-SoundFromWindTurbines-FS5-G.pdf). It is more difficult to respond to the issue of aesthetics. The experience with Cape Wind is a case in point.

The Cape Wind project would locate 130 3.6 MW turbines in a region of exceptional scenic beauty in Nantucket Sound, 3.5 miles off the coast of Cape Cod in Massachusetts. Opponents, including the late Senator Kennedy, current Secretary of State John Kerry, former Massachusetts governor and presidential candidate Mitt Romney, and conservative industrialist Bill Koch, argued that the turbines would negatively impact the scenic views from beaches accessible to the public and that their presence would reduce the value of expensive coastal property. After a long battle, the project was finally approved and, with financing arranged in 2014, is now due to begin construction. Opposition

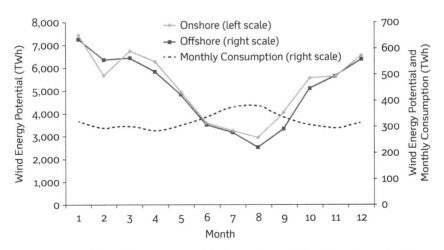

FIGURE 10.5 Monthly wind energy potential for the contiguous United States in 2006 and monthly electricity consumption for the entire United States.

to the installation of wind farms in the agricultural and ranching regions of the US Midwest has been more muted, with landowners receiving rents of as much as $6,000 annually for permission to site wind turbines on their property. And it should be noted that since the land footprint of the turbines is relatively small, one turbine for every 69 acres as indicated earlier, their presence has relatively little impact on conventional uses of the land.

CHALLENGES POSED BY THE VARIABILITY OF WIND

Operators of an electric utility face a formidable and continuing challenge to ensure that production of electricity is targeted in real time to meet projected demand. As discussed in Chapter 9, nuclear and coal-fired systems provide sources of what is referred to as baseload power. That is to say, the assumption is that these systems will operate essentially continuously, with minimal opportunity to respond to either increases or decreases in demand. Typically, gas-fired systems, which can be turned on or off rapidly, provide the flexibility needed to react to changes in demand. Accommodating an input of power from an intrinsically variable source such as wind poses a particular problem for the orderly operation of a complex electric utility network. The challenge is especially serious if wind is expected to make an important contribution (20% or more) to the overall demand for electricity.

Several approaches are available to accommodate variable inputs of power from an intermittent source such as wind. In the best of all worlds, one might hope for an opportunity to store power when it is available in excess of demand, to release it later as required. Pumped hydroelectric storage provides a potential opportunity to meet this objective. In this case, electricity, when available in excess of demand, is used to pump water uphill to a storage reservoir. When the demand resumes, the water can be released from the reservoir and allowed to flow downhill with the kinetic energy generated in the process employed to drive a turbine with consequent production of electricity. The overall efficiency of pumped hydroelectric systems can range as high as 80%. That is to say, 80% of the electricity consumed in pumping the water uphill can be recovered by tapping the energy available in the return flow. The balance, of course, is wasted.

The capacity for pumped hydroelectric facilities installed in the United States amounts to about 20 GW, capable of providing storage for approximately 2% of the nation's overall generating capacity. Pumped hydroelectric facilities are deployed in the United States primarily to minimize the difference between demand for electricity during peak and off-peak periods. When demand is low, at night, for example, and when wholesale prices for power are cheap, electricity can be used to pump water uphill. When demand is high and wholesale prices are elevated, during the following day, for example, water can be released from the storage reservoirs. The result is an increase in not only the economic but also the energy efficiency of the overall power system. With access to pumped hydroelectric facilities, baseload generating systems, coal-fired plants, for example, can be allowed to operate on a more or less continuous basis, eliminating the need for additional consumption of fuel (and related increases in CO_2 emissions) required to accommodate temporary shutdowns and restarts of these plants. Operators of pumped hydroelectric facilities profit by buying electricity when it is cheap, reconstituting it, and selling it subsequently when prices are more favorable.

Pumped hydroelectric facilities in the United States are located mainly in the eastern and western parts of the country where topographic conditions are favorable for their deployment (where it is hilly). As indicated in Figure 10.4, resources for onshore wind-generated power in the United States are concentrated to a large extent in the central part of the country where topographic relief is minimal. As a consequence, pumped hydroelectric facilities are unlikely to play a major role in the future development of onshore wind resources in the United States. They play an important role, however, in balancing supply and demand for power from wind in Europe. When power is available in excess of demand in Denmark, for example, it is transmitted as a matter of course to Norway to be processed for subsequent return through that country's extensive pumped hydroelectric facilities. A better option to accommodate variability of supplies associated with a major contribution from wind in the United States may be to couple outputs from wind farms distributed over an extended geographic region. And, as we shall see, there are comparable opportunities to benefit from this strategy in China.

The strength of the wind experienced at any particular location depends, particularly in winter, on the passage of large-scale weather systems across the country. When wind speeds are high at any particular location, they may be relatively slack in regions separated by 1,000 km or so, a rough estimate of the spatial scales associated with these meteorological disturbances. Archer and Jacobson (2007) considered advantages that could be realized by interconnecting wind farms over a region of 850 km by 850 km covering parts of Colorado, Kansas, Oklahoma, New Mexico, and Texas. They found that on average 33% of wind power from interconnected farms in this region could be exploited as reliable, baseload power. Opportunities available from coupling widely distributed, land-based, wind farms were discussed further by Katzenstein et al. (2010) and by Fertig et al. (2012). Kempton et al. (2010) drew attention to benefits that could be realized by connecting offshore wind farms distributed along the eastern seaboard of the United States. Lu et al. (2013) concluded that as much as 28% of the overall potential for offshore wind resources in China could be deployed as baseload power, cutting back significantly on the need for new, coal-fired plants that would be required otherwise to meet the increasing demand for electricity in that country's rapidly developing coastal provinces.

Meteorological factors responsible for the temporal variability of wind were discussed by Huang et al. (2014). They argued that the high-frequency variability of wind-generated power (variability on time scales as brief as minutes to hours) arises mainly as a result of locally generated small-scale turbulence. It could be eliminated effectively, they argued, by coupling outputs from as few as 5 to 10 wind farms distributed uniformly (equally separated) over a 10-state region of the central United States. More than 95% of the residual variability in the integrated power output would be expressed in this case on time scales longer than a day, providing an opportunity for grid operators to take advantage of credible multiday weather forecasts in scheduling anticipated future contributions from wind.

As indicated at the outset, the present electric system functions largely on a demand-driven basis. When consumers need electricity, they assume that it will be available. A more efficient system could incorporate a combination of demand- and supply-dictated considerations. If power is in short demand, utilities should be able to signal customers to reduce demand—to raise settings on their thermostats, for example, on a hot summer day or to postpone running electric clothes driers until later in the evening. Consumers could be encouraged to enter into such agreements with their utilities if they could see a clear benefit in terms of personal savings. With real-time pricing of power, consumers would be encouraged to use electricity when it is abundant and cheap, and to conserve it when it is scarce and expensive. In the process, utilities would be better able to manage the challenges posed by fluctuations in supply from variable sources of power such as wind.

Advances in battery technology could provide an additional opportunity to address the potential source/demand power mismatch issue. In a recent paper, Huskinson et al. (2014) described a prototype of a flow battery, based on the oxidation and reduction of readily available small organic molecules. Think of two large tanks of chemicals that could be connected to either store or produce electricity. And imagine that the tanks could be large enough to accommodate utility-level demands for storage. This could have dramatic benefits for the renewable energy industry. Whether it can meet this lofty expectation will depend, however, not only on the success of future advances in technology but also ultimately on the cost for deployment of this technology: absent government subsidies, incentives from the profit motive will prevail.

Earlier in this volume, we made the point that there were two requirements to meeting the objective of reducing emissions of CO_2 from an economy such as the United States. First, we will need to transition to an economy where electricity is produced primarily from sources with minimal emission of CO_2 to the atmosphere. Second, we need to develop a source for energy in the transportation sector that does not depend exclusively on oil. As will be discussed later, if we can find a way to power our cars and trucks with electricity, and if these vehicles are equipped with batteries that can be charged from the grid with low-carbon-emitting sources of electricity, we can make significant progress in addressing the second of these objectives. If utilities should have the options to draw power from the batteries of these vehicles when demand exceeds supply, this could contribute to a win-win option for both suppliers and consumers of electric power. As an important byproduct, it could contribute to a more effective inclusion of power from wind and other intrinsically variable sources (such as solar discussed in the following chapter) in the overall electricity supply system.

ECONOMIC INCENTIVES FOR DEVELOPMENT OF WIND AND OTHER LOW-CARBON ENERGY SOURCES

The key incentive for development of wind power in the United States involves what is known as the production tax credit (PTC). As the name suggests, the PTC provides a tax rebate for producers of electricity from wind, with credits extending typically for 10 years following the time when the wind facility first goes into service. The PTC incentive has been allowed to expire five times and has been extended five times since it was first introduced in 1992. The result has been a boom-bust cycle for wind investments in the United States, as illustrated in Figure 10.6. As indicated, the pace of investments in wind systems drops off significantly in years following expiration of the PTC. Uncertainty as to whether the authorization in place for 2012 would be extended through 2013 resulted, as indicated in the figure, in a precipitous drop in new starts for wind projects in 2013. Installations fell to a level as low as 70.6 MW during the first three quarters of 2013, a decrease of 96% with respect to the record-setting preceding year. As it turned out, the PTC was extended in January 2013 after a brief lapse. The new bill provided an extension of the PTC for wind projects on which construction began prior to January 1, 2014. This spurred a resurgence in investment, resulting in a significant increase in capacity scheduled to come online in 2014 (1 GW as of the first half of 2014 with a further 14 GW under construction).

The American Reinvestment and Recovery Act (ARRA) of 2009 provided incentives for investments in renewable energy in addition to the PTC, an investment tax credit (ITC), for example, that could be applied as a substitute for the PTC. As a further

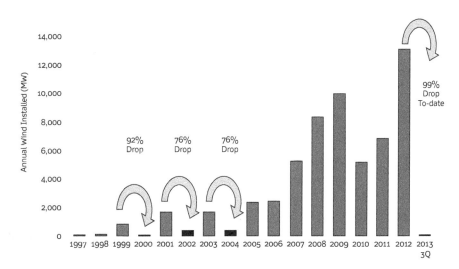

FIGURE 10.6 Annual installations of wind-generated electricity in the United States (in MW) indicating the significant decrease in investments when continuity of the PTC was in question, notably in 2000, 2002, 2004, and 2013.
(*Source:* American Wind Energy Association 2014, https://www.awea.org/Advocacy/Content.aspx?ItemNumber=797)

optional incentive, ARRA decreed that wind projects initiated in 2009 and 2010 would be eligible for grants from the Treasury Department ranging up to 30% of the cost for development of specific projects. As with the PTC, these opportunities would be available only through the end of calendar year 2012. Clearly, improved certainty with respect to the future of policy initiatives such as the PTC and ITC would be helpful. A more sustained commitment, even with a declining standard for related subsidies, could contribute to a more orderly, and ultimately a more successful development of wind resources in the United States.

Some 30 states in the United States have adopted what are referred to as renewable portfolio standards (RPSs) as state-level incentives for investments in renewable energy. Typically, RPSs require electricity suppliers to include a specified minimum fraction of renewable energy in the mix of power they deliver to consumers. Requirements differ from state to state. Massachusetts, for example, specifies that 15% of electricity consumed in 2020 should be derived from renewable sources. It requires further that the fraction of renewables in the mix should increase by 1% for every year thereafter—or until the Massachusetts legislature chooses to adjust this mandate. California's requirements are even more impressive: 33% renewables by 2020.

Renewable sources of electricity such as wind, solar, and hydro differ from conventional sources such as coal, oil, natural gas, and nuclear in that fuel in the former case is effectively free (you don't have to pay for the wind or the sun or the rain that falls from the sky). Investors in renewable sources can make a risk-free decision if they know precisely what to expect by way of income from selling the power they produce: their expense is determined almost totally by the initial outlay of capital. The feed-in tariff (FIT) approach favored in a number of countries guarantees investors a specific price for their electricity over essentially the lifetime of their investment. This approach has been spectacularly successful in stimulating development of renewable energy, most notably in Germany.

Germany adopted an aggressive position with respect to its ambitions to replace climate-impacting fossil fuels with low to zero carbon-emitting renewable alternatives. German's current plans call for renewable sources to account for 35% of electricity production by 2020, 50% by 2030, and a minimum of 80% by 2050. The FIT approach is the policy option of choice to advance this agenda. Investors are guaranteed a specific price for the power they produce over a 20-year period, the projected operational lifetime of their investments. Prices differ for different renewable options and are designed to encourage a range of technological choices. FITs for roof-mounted solar arrays, for example, are higher than those for onshore wind farms, reflecting the higher levelized costs projected for the former. Prices are set to provide investors with a reasonable return on their investments, typically between 5% and 7%. Terms for prospective new investments are adjusted on a more or less annual basis to reflect changes in costs for specific renewable opportunities. Utilities are required to purchase the power produced from renewable sources according to the original contracted FIT price for this power. Prices incurred in purchasing power under the FITs are passed on to the consumer. The danger with this system is that these expenses could become prohibitively high. But, to date, this appears not to have been the case for Germany. The FIT is estimated to account for about 18% of present retail electricity prices. Renewable energy currently accounts for 22% of German electricity. The advantage of the FIT approach as practiced in Germany is that the government can influence the pace of renewable development by adjusting the FIT as required (Paris 2009).

The Chinese response to encouraging the development of renewable energy, specifically wind-generated electricity, differs somewhat from the approaches adopted in Germany and in the United States. China chose to encourage wind development in two phases. In phase 1, from 2003 to 2008, companies were invited to propose to develop wind farms in specifically designated areas of the country with the grid company assigned responsibility (and the expense) for connecting output from these facilities to the grid. Proposals would be vetted based in part on what they offered by way of revenue expected per unit of electricity generated over (approximately) the first 10 years of operation of the facilities. China switched more recently to initiatives that have more in common with the policies favored in Germany: essentially FIT with government-defined revenues guaranteed for specific government-identified opportunities for specific renewable energy developments.

KEY POINTS

1. China ranks number 1 in the world in terms of installed wind capacity but trails the United States in related production of electricity, reflecting requirements in China for wind facilities to be shut down frequently in winter when coal-fired combined heat and power plants are obliged to operate to supply hot water for district heating. Strategies to use electricity rather than hot water as the energy source for heating buildings could alleviate this inefficiency.
2. Wind accounted for 3.5% of electricity produced in the United States in 2012, 1.6% of electricity consumed in China in 2011.
3. Wind resources are sufficient in principle to provide the bulk of the electricity required to supply the needs of the major CO_2-emitting countries.
4. The variability of wind poses a challenge for utilities as they seek to reconcile supplies and demand for electric power. The challenge can be mitigated by coupling outputs from wind farms distributed over geographic regions separated by distances of up to 1,000 km.
5. The electricity economy operates currently on a demand-driven basis. It could function more efficiently if utilities had the ability to adjust demand for non-essential electricity-powered services when supplies are limited. Elements of

what is referred to as smart grid technology could make it easier for utilities to incorporate variable sources such as wind in their composite power mix with related improvements in the overall efficiency of the power system.

6. A range of incentives is available in different countries to encourage development of renewable sources of energy. Federal production tax credits, investment tax credits, and state incentives are important in the United States but suffer in their effectiveness due to a lack of long-term commitment.

7. The goal in Germany is for renewable sources of energy to account for a minimum of 80% of electric power supply by 2050. A combination of wind, solar, and other renewable sources could be comparably effective in the United States. Accomplishing such an objective would require credible, durable, commitments from federal, state, and local authorities.

8. Transitioning from fossil to renewable sources of energy would allow for important visibility and stability with respect to future prices for electricity. Levelized costs in the latter case would be determined primarily by the upfront costs for capital. Uncertainties in future prices for fuel make it more difficult to determine levelized costs for conventional coal-, gas-, and oil-fired systems.

Notes

1. Predictions of future weather are based on complex computer models initiated using a variety of sources of data—measurements from ground-based stations, aircraft, balloons, ships, ocean buoys, and satellites. As we know from experience, these predictions are reasonably reliable for forecast periods up to as much as a week. If the composite forecast for Boston tells us to expect snow in the afternoon tomorrow, we can pretty well make plans to adjust accordingly. The reanalysis technique employs computer models similar to those used for weather forecasts. The reanalysis approach has the advantage that it knows what the answer should be based on the historical record. The data input to the simulation procedure can be refined accordingly. Rather than running the model for 10 days or so in the future, as is the practice for forecast purposes, the reanalysis simulation is updated typically every 6 hours to incorporate the historical data. The reanalysis approach, in principal, should provide the best possible record of past changes in the properties of the atmosphere.

2. The capacity factor (CF) is a measure of the fraction of the rated potential of a specific power-generating facility that the facility is able to realize over a representative operational year. Restricting the analysis to CF values in excess of 20% presumes that it would make little economic sense to locate expensive wind turbines in environments where wind conditions would be insufficient to meet this limit.

3. PWh identifies an energy unit corresponding to an output of 10^{15} watts of power (peta means 10^{15} or 1,000 trillion) for an hour. An energy output of 1 PWh is equivalent to an energy content of 3.4095 quad. Recall from Chapter 2 that the quad (10^{15} BTU) is the unit of energy most commonly used to report data on national energy consumption. Global consumption of energy in all forms registered 524 quad in 2010 with 69 quad supplied in the form of electricity.

4. The tabulation from the original publication has been updated to rank countries based on emissions for 2012 rather than 2005, the year selected for focus in the original publication. The revised table has been updated also to account for 2011 data for national-level consumption of electricity. The United States ranked number 1 in terms of emissions in 2005, supplanted since by China. Demand for electricity until recently was highest in the United States, but the difference between the United States and China has narrowed significantly in the interim, to the point where consumption of electricity in China in 2011 now exceeds that in the United States by 325 TWh.

References

Archer, C. L., and M. Z. Jacobson. 2007. Supplying baseload power and reducing transmission requirements by interconnecting wind farms. *Journal of Applied Meteorology and Climatology* 46: 1701–1717.

Chen, X., X. Lu, M. B. McElroy, C. P. Nielsen, and C. Kang. 2014. Synergies of wind power and electrified space heating: Case study for Beijing. *Environmental Science and Technology* 48, no. 3: 2016–2024.

GE. 2006. *2.5 MW series wind turbine*. Fairfield, CT: General Electric Energy.

GWEC. 2013. *Global wind report: Annual market update 2012*. Brussels, Belgium: Global Wind Energy Council.

Huang, J., X. Lu, and M. B. McElroy. 2014. Meteorologically defined limits to reduction in the variability of outputs from a coupled wind farm system in the Central US. *Renewable Energy* 62: 331–340.

Huskinson, B., M. P. Marshak, C. Suh, S. Er, M. R. Gerhardt, C. J. Galvin, X. Chen, A. Aspuru-Guzik, R. G. Gordon, and M. J. Aziz. 2014. A metal-free organic-inorganic aqueous flow battery. *Nature* 505: 195–198.

Katzenstein, W., E. Fertig, and J. Apt. 2010. The variability of interconnected wind plants. *Energy Policy* 38: 4400–4410.

Kempton, W., F. M. Pimenta, D. E. Veron, and B. A. Colle. 2010. Electric power from offshore wind via synoptic-scale interconnection. *Proceedings of the National Academy of Sciences USA* 107: 7240–7245.

Lu, X., M. B. McElroy, and J. Kiviluoma. 2009. Global potential for wind-generated electricity. *Proceedings of the National Academy of Sciences USA* 106: 10933–10938.

Lu, X., M. B. McElroy, C. P. Nielsen, X. Chen, and J. Huang. 2013. Optimal integration of offshore wind power for a steadier, environmentally friendlier supply of electricity in China. *Energy Policy* 62: 131–138.

Masters, G. M. 2004. *Renewable and efficient electric power systems*. Hoboken, NJ: John Wiley & Sons, Inc.

Paris, J. A. 2009. The cost of wind, the price of wind, the value of wind. *European Tribune*, May 6, 2009.

11

Power from the Sun

ABUNDANT BUT EXPENSIVE

AS DISCUSSED IN the preceding chapter, wind resources available from nonforested, nonurban, land-based environments in the United States are more than sufficient to meet present and projected future US demand for electricity. Wind resources are comparably abundant elsewhere. As indicated in Table 10.2, a combination of onshore and offshore wind could accommodate prospective demand for electricity for all of the countries classified as top-10 emitters of CO_2.

Solar energy reaching the Earth's surface averages about 200 W m^{-2} (Fig. 4.1). If this power source could be converted to electricity with an efficiency of 20%, as little as 0.1% of the land area of the United States (3% of the area of Arizona) could supply the bulk of US demand for electricity. As discussed later in this chapter, the potential source of power from the sun is significant even for sun-deprived countries such as Germany.

Wind and solar energy provide potentially complementary sources of electricity in the sense that when the supply from one is low, there is a good chance that it may be offset by a higher contribution from the other. Winds blow strongest typically at night and in winter. The potential supply of energy from the sun, in contrast, is highest during the day and in summer. The source from the sun is better matched thus than wind to respond to the seasonal pattern of demand for electricity, at least for the United States (as indicated in Fig. 10.5).

There are two approaches available to convert energy from the sun to electricity. The first involves using photovoltaic (PV) cells, devices in which absorption of radiation results directly in production of electricity. The second is less direct. It requires solar energy to be captured and deployed first to produce heat, with the heat used subsequently to generate steam, the steam applied then to drive a turbine. The sequence in this case is similar to that used to generate electricity in conventional coal, oil, natural gas, and nuclear-powered systems. The difference is that the energy source is light from the sun rather than a carbon-based fossil fuel or fissionable uranium. As discussed later, sunlight in this second approach is normally concentrated to enhance its overall efficiency. This technology is referred to as concentrated solar power (CSP).

The chapter begins with a review of the physical principles that underscore the PV option. It discusses the steps involved from refining of the silicate minerals used to produce the PV material in the first place, to assembly of the material into panels, and to its eventual deployment on rooftops, in ground-mounted systems, or as components of utility-scale power-generating facilities. It continues with a discussion of the underlying economics. What does it cost to manufacture and install PV systems? How have these costs varied over time, and what are the prospects for the future? As discussed in Chapter 9, including note 4, the key parameter defining the economic viability of any particular electric power investment is what is known as the levelized cost. What are the levelized costs for different PV systems (residential, commercial, and utility)? The chapter continues with a discussion of the current status and prospects for different forms of CSP, concluding with a summary of key points.

HOW PHOTOVOLTAIC SYSTEMS CONVERT LIGHT TO ELECTRICITY

The basic component of a typical PV system is a cell composed of crystalline silicon. The silicon atom in its electrically neutral state includes 14 negatively charged electrons orbiting a nucleus defined by 14 positively charged protons. Ten of the electrons are tightly bound to the nucleus. The remaining four, referred to as valence electrons, are less strongly constrained. They have the ability to combine with valence electrons from other atoms to provide the glue that allows individual atoms to link up with neighbors to form more complex structures. Silicon atoms are connected in this case by what are referred to as covalent bonds contributed by pairs of shared electrons.[1] The result is the formation of a crystal in which each individual silicon atom is linked covalently to four neighbors.

Silicon in crystalline form is classified as a semiconductor. That is to say, it can conduct electricity but not very well. What determines the conductivity of a substance is the concentration of free electrons, the number of electrons present in what is referred to as the conduction band. A metal such as copper has many free electrons. Silicon has very few except at high temperature. For silicon to behave more like a conductor, energy must be communicated to electrons, prompting them to raise their energy state in order to populate the conduction band. In a PV cell, the energy to accomplish this objective is provided by the absorption of radiation—light in the near infrared and visible portions of the electromagnetic spectrum. The energy required for an electron to transition to the conduction band is referred to as the band gap energy. For silicon this amounts to 1.12

electron volts (eV), corresponding to a wavelength threshold of 1.11 micrometers (μm).[2] If the energy supplied is greater than 1.12 eV (light with wavelength less than 1.11 μm), the excess is communicated to the semiconductor in the form of heat.

When an electron is released from an atom to populate the conduction band, it leaves behind a positively charged partner referred to as a hole. In the absence of an electric field to separate electrons and holes—to keep them apart—this process is likely to reverse; the electrons and holes can reconnect to restore charge neutrality. Light with wavelength equivalent to the energy of the band gap is emitted in this case. This sequence provides the basis for the operation of light-emitting diodes (LEDs) employed in a variety of modern applications, including energy-efficient, long-lasting light bulbs. Before the crystalline silicon can operate as a PV system, it must be doped with the addition of small concentrations of impurities and separated into distinct units referred to as n-type and p-type materials. The electric field required for the silicon to operate as a PV system arises as a result of a separation of charge that develops at the junction of these materials when they are brought into contact.

Adding phosphorus even at a concentration as low as one part in a thousand leads to an important increase in the capacity of a silicon crystal to produce free electrons. The phosphorus atom has five valence electrons, one more than silicon. When the phosphorus atoms bind to silicon in the modified crystal, electrons are released and are consequently free to migrate. This defines the important distinctive property of an n-type semiconductor.

Precisely the opposite situation applies with p-type semiconductors. The silicon is doped in this case with atoms defined by one fewer valence electron than silicon. A common choice, with three valence electrons, is boron. When a boron atom binds to four silicon neighbors in the modified crystal, it has to borrow an electron from a silicon atom to complete the process. This results in formation of a negatively charged unit centered on the boron atom, offset by a positively charged hole associated with the silicon atom that provided the electron required to stabilize the boron in the first place. The positively charged silicon atom can borrow an electron subsequently from a silicon neighbor in a process that can be repeated many times, allowing holes to migrate freely throughout the crystal. This defines the essential property of a p-type semiconductor.

When p and n units are brought together, there develops not surprisingly an excess of electrons initially on the n side of the juncture, a surplus of holes on the p side. Electrons can drift across the interface, filling holes on the p side, leaving holes behind on the n side. This leads to a separation of charge with consequent production of an electric field in what is referred to as the depletion region. The electric field is reflected in a drop in voltage of approximately 0.5 V from the n to the p side of the juncture. The depletion region can be exceptionally thin, measuring as little as 1 micrometer (μm) in thickness. Its formation and persistence is critical, however, to the operation of the n-p composite as a PV electricity-generating system.

If the p-n combination is exposed to light and if the n and p components are linked with a conducting wire, the cell can provide a source of low-voltage electricity. The basic structure of an n-p cell is illustrated in Figure 11.1. The voltage can be increased by connecting cells in series to form what are referred to as modules. A typical example is displayed in Figure 11.2.

STEPS INVOLVED IN PRODUCING SILICON-BASED PHOTOVOLTAIC SYSTEMS

Silicon, present primarily in the form of silicate minerals (SiO_2), is the second most abundant element in the Earth's crust. As discussed by Masters (2004), a complex series of transformations is involved in converting the original silicon feedstock to the form in

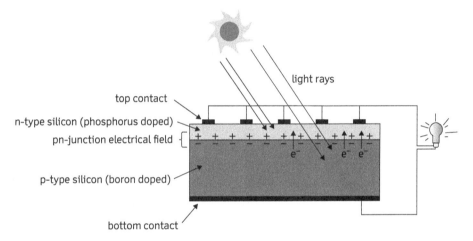

FIGURE 11.1 Basic structure of an n-p cell.
(*Source:* http://www.lrc.rpi.edu/programs/nlpip/lightingAnswers/photovoltaic/04-photovoltaic-panels-work.asp)

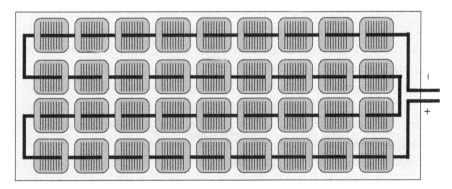

FIGURE 11.2 Illustration of a photovoltaic module with 36 cells connected in series.
(*Source:* http://pvcdrom.pveducation.org/MODULE/Design.htm)

which it is incorporated in the modules that constitute the operational components of a PV system.

The first energy-intensive step involves purification and treatment of the silicon to produce liquid trichlorosilane ($SiHCl_3$). The trichlorosilane is allowed then to react with hydrogen (H_2) in a high-temperature vat to form essentially pure silicon. From a chemical viewpoint the overall process involves what is referred to as reduction of the silicon, its transformation from the oxidized form in which it is present as SiO_2 to the reduced form represented by pure Si.[3] The rock-like chunks of essentially pure silicon produced in this process are heated subsequently to temperatures in excess of 1,400°C (Masters 2004). Small concentrations of impurities can be introduced at this stage to allow for eventual production of either p- or n-type materials. Introducing small seeds of solid crystalline silicon (pencil size as described by Masters) into the melt provides a substrate on which the molten silicon can condense to form solid crystalline silicon (with or without the doped additions). As the seeds are rotated and withdrawn from the melt, they provide a source of essentially pure solid crystalline ingots, each measuring approximately 1 meter in length, 20 centimeters in diameter. The ingots are then sawed and shaped into thin rectangular units referred to as wafers. This allows for a higher packing density of the silicon wafers when they are assembled into PV modules. A significant fraction of the silicon is lost in the process, as much as 50%, the silicon analogue of the sawdust waste produced in working with wood.

In practice, as illustrated in Figure 11.1, the p-type component constitutes the bulk of the material incorporated in a typical PV cell. The n-type junction may be formed by exposing the top of the p-type material to a sufficient concentration of phosphorus to override the presence of boron in a surficial layer. The result is production of what is basically a p-type semiconductor overlain by a thin layer of n-type material providing the junction needed to exploit the PV function of the composite.

A combination of 36 cells can furnish power at 12 V (actually a little more if required). Modules incorporating up to 72 cells are available, providing power at a voltage of up to 24 V or higher. A typical cell is less than 500 micrometers (μm) thick (0.02 inches) with lateral dimensions of approximately 15 x 15 cm. Modules weigh between 34 and 62 pounds and are produced by connecting individual cells on a plastic substrate mounted on an aluminum frame. An antireflective material is applied on the top to protect the module from the external environment, serving at the same time as a means to enhance the flux of sunlight reaching the PV material. A glass composed of tin oxide is effective in transmitting red, green, and yellow light, reflecting light in the blue region of the spectrum, accounting for the characteristic blue color of many of the modules in common use.

Modules are interconnected in patterns, referred to as arrays, designed to provide the required combination of voltage and current. Coupling components in series increases the resulting voltage; coupling in parallel contributes to an increase in current. The product of current and voltage determines the net power output of the system, constrained ultimately by the flux of sunlight intercepted by the PV array.[4]

CURRENT STATUS OF PHOTOVOLTAIC ENERGY

PV has enjoyed a period of remarkable growth worldwide over the past decade, from 2.2 GW of installed capacity in 2002 to close to 100 GW in 2012. Additions in 2012 alone amounted to 30.5 GW. Despite this growth, PV continues to represent a small fraction of total global electricity-generating capacity, approximately 2% in 2012. By way of comparison, the contribution from wind, onshore and offshore, amounted to 5% of global capacity in 2012. Germany tops the list in terms of PV installations with 32% of total global capacity in 2012, followed by Italy with 16%, the United States with 7.2%, China with 7%, and Japan with 6.6%. Developments in the European Union (EU) have been particularly impressive. Including Germany, the EU accounted for 53.2% of the 2012 total installed global PV capacity.

Bloomberg New Energy Finance (2013) predicts that PV will grow to 16% of the global total by 2030, comparable to the level (17%) projected for wind. Government incentives, notably the feed-in tariff initiatives favored in Europe, together with comparable measures adopted in the United States, China, and Japan, have played an important role in stimulating much of the recent growth. The question is whether the momentum experienced in the past can be maintained in the face of likely reductions in the scale of incentives in the future. If not, the projections by Bloomberg may prove to be optimistic.

As indicated in Figure 11.3, installed prices for small residential and commercial PV systems (capacities less than 10kW) decreased in the United States from $12 per watt in 1998 to $5.3 per watt in 2012. Much of this change developed over the past 5 years, attributed primarily to the precipitous recent drop in prices for the PV modules. The cost for modules in 2007 averaged approximately $2 per watt. By 2012, the price had dropped to less than 80 cents per watt. An important ramp-up in production in China was largely responsible for this decrease. Chinese companies accounted for 30.6% of total global production in 2012, with the bulk of the product directed toward markets overseas.

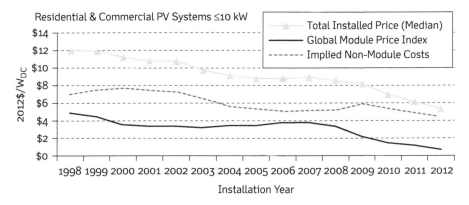

FIGURE 11.3 Installed prices for small residential and commercial photovoltaic systems in the United States.
(*Source:* Barbose et al. 2013)

The dominance of Chinese companies in the market for PV systems (both cells and modules) has triggered a series of complaints from both the United States and the EU to the effect that China is violating international trade norms by dumping PV products at below cost on to both the United States and EU markets. China has responded by accusing the United States and the EU of violating the same norms by subsidizing exports of polysilicon, the feedstock for production of PV systems. As of April 2014, these disputes had not been resolved. In the meantime, wholesale prices for PV cells and modules have continued to fall.

The cost for installation of a 2–5 kW PV system in Germany is approximately half (on a per watt basis) of what it is in the United States: $2.60/W as compared to $5.20/W. Only in Japan are costs ($5.90/W) higher than they are in the United States (Barbose et al. 2013). Nonmodule costs include expenses for the hardware needed to mount the PV panels on customer roofs, costs for the ancillary equipment such as the inverters needed to convert the direct current (DC) output of PV to the alternating current supply (AC) required by both households and the grid, costs for labor, and a variety of what are identified as soft costs adding up to a total of $3.30/W. A breakdown of total non-module costs for the United States is displayed in Figure 11.4 (Friedman et al. 2013).

The SunShot initiative of the US Department of Energy (http://energy.gov/eere/sunshot/about) aims to reduce the contribution of soft costs for residential systems in the United States to $0.65/W by 2020, to $0.44/W for commercial systems. The overall target is to lower total installed costs to $1.50/W for residences and $1.25/W for businesses.

Reasons for the difference between soft costs for PV systems in the United States and Germany are discussed by Seel et al. (2013). They attribute part of the difference to the

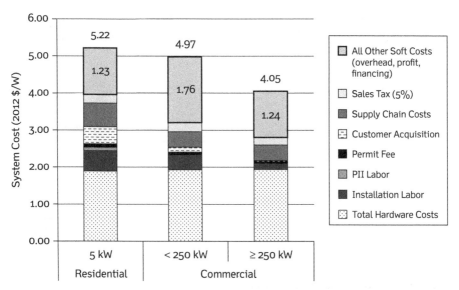

FIGURE 11.4 A breakdown of module costs for the United States in the first half of 2012. (*Source:* Friedman et al. 2013)

more developed and consequently more competitive environment for PV in Germany. Other considerations include the fact that companies in the PV installation business in the United States have to expend more capital on promotion and advertising to acquire their customers than do their counterparts in Germany. Further, labor costs for installation are higher in the United States. And, finally, procedures involved in applications, permitting, and inspections for PV installations are more complex in the United States. The expectation in the SunShot initiative is that with increasing experience with PV in the United States, costs for installations in the two countries should begin to converge. Expansion of the market for PV in the United States is expected to increase customer awareness of opportunities and at the same time provide a stimulus for development of an expanded, qualified, and more competitive labor pool. Simplifying procedures for permitting and fast-tracking approvals for PV installations in the United States may prove challenging, however, given the diversity of governmental bodies (local, regional, state, and federal) with responsibility for oversight in the United States as compared with the more streamlined structure in place in Germany.

Assuming the current average price of $5.20/W for installation of residential PV systems in the United States, assuming further a capacity factor for operation of these systems of 20% (the actual power delivered as a percentage of the nameplate potential maximum), an operational lifetime of 20 years and a discount rate of 7% (essentially a measure of the cost of capital), the levelized cost (present dollars) for electricity required to support a current residential investment in PV in the United States is estimated at 26.2 cents/kWh. This compares unfavorably with levelized costs for alternative investments as summarized in Table 9.2: the cost for onshore wind, for example, is almost a factor of 3 lower at 8.7 cents/kWh. The lower installation cost for PV systems in Germany is consistent with a levelized cost for residential systems in that country of 13.1 cents/kWh. If we assume success in meeting the optimistic SunShot price objectives for the United States in 2020—$1.50/W for residential PV systems, $1.25/W for commercial systems—imputed levelized costs would drop in this case to 7.6 cents/kWh and 6.3 cents/kWh, respectively. Investments in PV systems would be competitive economically then with onshore wind. Levelized costs for large-scale facilities (capacities greater than those associated with residential investments) may be derived by scaling costs for installation based, for example, on the data presented in Figure 11.4.

Given the relatively high levelized cost for PV systems compared with alternatives such as wind, it may be difficult to understand the success the PV industry has enjoyed worldwide in recent years. The drop in prices for modules was surely a contributing factor. Access to generous feed-in tariffs was also important. A key factor, however, accounting at least in part for the success of the solar industry in the United States, has been the emergence of third-party brokers, companies serving as intermediaries between households interested in installing PV systems but reluctant to pay the upfront cost, and utilities seeking to acquire at least a portion of the resulting power. I can tell the story from personal experience of how this works.

Our family has a summer home on Cape Cod. The house, equipped with air conditioning, is occupied mainly in the summer. The average annual consumption of electricity amounts to 5,064 kWh. Retail prices for electricity on Cape Cod are high, 26.0 cents/kWh, with more than half of the cost assigned to delivery rather than production. We decided a year or so ago to explore the possibility of installing PV panels on our house and contacted a company, Solar City, in the PV business to explore options. They concluded that a 5.15 kW DC PV system could supply a quantity of electricity over the course of a year equal to our present annual consumption. The electricity generated by this system would exceed household demand over approximately half of the year with an offsetting shortfall in the remainder. The strategy was that the house should continue to be connected to the grid. It would deliver power to the grid when production exceeded demand, reversing the direction of flow when supply was insufficient.

The cost for installing the system was estimated at $25,725. With an upfront rebate of $2,000, the out-of-pocket immediate expense would amount then to $23,725. Over time (20 years), if we owned the system, we could expect to further reduce the net cost by taking advantage of various federal and state tax incentives and by selling power to the grid under the Solar Renewable Energy Certificate (SREC) program available in Massachusetts. The projection was that should we choose to purchase the system and to have the company install and manage it, the investment would be profitable after 10 years (a 10-year payback).

We chose not to accept this proposal. We elected rather to proceed with a second option, one in which Solar City would finance the system installed on our roof. They would own it and would be totally responsible for its maintenance and for monitoring its performance. We would pay an upfront fee of $2,000. In return we would be guaranteed a fixed price of 13.1 cents/kWh for our electricity over a 20-year period, half of what we were paying previously. The house would continue to be connected to the grid and we would pay a monthly fee of $6.50 to the utility for this privilege. How could Solar City afford to offer this option? Clearly, they would be better organized than we to take advantage of all of the available rebate and tax incentives. They could package the power from multiple installations and market it in the form of options for future delivery under the SREC program. And they would be transferring power to the grid at the prevailing retail price (26 cents/kWh at present) by running our meter in reverse. The arrangement was obviously good and profitable for Solar City but also attractive for us, an example of the critical supportive role third parties are currently playing in the PV business in the United States. Not surprisingly, third-party arrangements accounted for close to 60% of all PV installations in the United States in 2012.

A key question is whether the current financial arrangements are sustainable. They work at present in part because the footprint for home-based PV systems is relatively small in the context of the overall electricity power system. Particularly supportive is the ability of the third parties to sell power to the grid at the prevailing retail rate. Were they required to sell at the wholesale rather than retail price, given the high levelized present cost for PV, the situation would be very different. Utilities may be expected to argue in the future that since they provide and maintain the infrastructure that connects homes to the grid, they should be compensated accordingly. This issue has already been raised in Arizona, where the installation of home-based PV is significantly higher than it is in Massachusetts. A further question concerns the long-term future of the additional incentives. Despite the uncertainties raised by these considerations, companies such as Solar City clearly see a future, and Wall Street seems to agree. Since Solar City went public in December 2012 with an initial offering price for its stock of $10, the price of its shares has surged, to a high of $83 in early April of 2014 before settling back to about $50 in late 2014, closer to $30 in early 2016.

As indicated earlier, the feed-in tariff option has played an important role in the success of PV in Germany. This strategy has been consequential also for China. A uniform feed-in tariff of 1.15 RMB/kWh (18.4 cents) was approved in China in July 2011 for projects operational prior to December 31, 2011. The problem that developed under this arrangement was that the bulk of the investments that ensued was concentrated in regions where solar conditions were most favorable, mainly in the west, rather than in regions where demand was highest, primarily in the east. The ground rules were amended subsequently to differentiate between regions depending on their solar resources, with three different levels of tariffs: 0.90 RMB/kWh (14.4 cents) in regions where solar conditions were most favorable; 0.95 RMB/kWh (15.2 cents) in regions where conditions were classified as intermediate; and 1.00 RMB/kWh (16 cents) in regions where conditions were less than optimal. Attesting to the significance of the feed-in tariff incentive, installations of PV systems in China increased from 8.3 GW in 2012 (27.2% of the global total) to 20.3 GW in 2013 (47.5% of the global total).

UTILITY-SCALE PHOTOVOLTAIC SOLAR POWER

For present purposes we define utility-scale installations as facilities with capacities greater than 10 MW. As of the end of 2012, the combined capacity of worldwide utility-scale PV amounted to 9.38 GW, approximately 9% of the total installed global PV base. Germany accounted for 30% of the utility scale total, followed by the United States with about 21%. Utility-scale investments, however, are increasing rapidly at the present time. Additions recorded in 2012 were responsible for an increase of 60% in the total installed global base. Investments in smaller scale household and commercial systems, however, continue to dominate the overall market.

Table 11.1 provides a summary of the 12 largest PV utility plants operational in the United States as of April 2012 (plants with capacities greater than or equal to 20 MW). The table includes also information on the developers of these plants, the years in which the plants went operational, the PV technologies they deployed, their locations, and the purchasing power agreements (PPAs) that supported their

TABLE 11.1

A Summary of the 12 Largest Photovoltaic Utility Plants Operational in the United States as of April 2012

Plant	MW	Developer	Operating Year	Solar Technology	Location	PPA with
Copper Mountain	48	Sempra Generation	2010	CdTe	Boulder City, Nevada	PG&E
Avenal Solar Generating Facility	45	NRG	2011	a-Si	Avenal, California	PG&E
Mesquite Solar 1	42	Sempra Generation	2011	c-Si	Arlington, Arizona	PG&E
Long Island Solar Farm LLC	32	BP Solar	2011	c-Si	Upton, New York	Long Island Power Authority
Cimarron I	30	First Solar	2010	CdTe	Cimarron, New Mexico	Tristate Generation and Transmission
FRV Webberville Plant	30	Fotowatio Renewable	2011	c-Si	Webberville, Texas	Austin Energy
San Luis Valley Solar Ranch	30	Iberdrola	2011	c-Si	Alamosa County, Colorado	Xcel
Agua Calliente (partial output)	30	First Solar	2012	CdTe	Yuma County, Arizona	MidAmerican Energy Holdings (Buffett)
Desoto Solar Energy	25	SunPower	2009	c-Si	Arcadia, Florida	FPL
Blythe Generating Facility	21	First Solar/NRG	2009	CdTe	Blythe, California	SCE
Road Runner Solar Electric Facility	20	NRG Energy	2011	CdTe	Santa Teresa, New Mexico	El Paso Electric
Stroud Solar Station	20	Cupertino Electric	2011	c-Si	Helm, California	PG&E

Source: Mendelson et al. (2012).

development (McGinn et al. 2013).[5] Three of the plants are located in California, two each in New Mexico and Arizona, with single facilities in Nevada, Texas, Florida, Colorado, and New York. With the exception of the plant in New York, all of these installations are sited in regions where solar conditions are optimal. Five of the plants use PV systems based on thin film cadmium telluride (CdTe) technology supplied by the First Solar Company. Crystalline silicon (c-Si) represents the technology of choice for six of the remaining seven plants included in the table, amorphous silicon (a-Si) for the seventh.

As of April 2012, 1,329.5 MW of utility-scale PV plants with capacities in the range of 20–49 MW were under construction in the United States (Mendelson et al. 2012). PPAs had been signed for a further 9,425 MWs with capacities greater than 50 MW. At the time this was written (late 2014), the Topaz Solar Farm in San Luis Obispo, California, owned by Warren Buffett's MidAmerican Energy Company, was operating at a capacity of 300 MW, projected to increase to 550 MW by 2015. A comparably large facility, Desert Sunlight, also 550 MW, is under construction in Riverside County, California. The Topaz facility is supported by a 25-year PPA with the California-based Pacific Gas and Electric Company (PG&E). Desert Sunlight is supported by two PPA agreements, one with PG&E and the other with Southern California Edison.

It comes as little surprise that California has been the focus for much of the recent development of solar power in the United States. California is legislatively committed to the requirement that 33% of the electricity consumed in the state in 2020 must be supplied by renewable sources. Renewables accounted for 15.4% of electricity consumed in 2012, composed as follows: 6.3% wind, 4.4% geothermal, 2.3% biomass, 1.5% small hydro, with solar accounting for only 0.9%. There is clearly scope for further growth of solar.

UTILITY-SCALE CONCENTRATED SOLAR POWER

CSP refers to an approach in which mirrors are used to focus sunlight on a target, where it can be used to heat a fluid with the energy concentrated in this manner deployed subsequently to generate electricity using a conventional steam turbine. In one design, the light is focused on a receiver sited on the top of a tower in the middle of the mirror array. In another, mirrors are deployed in parabolic-shaped, linear arrays. The light reflected by the mirrors is focused in this case to heat a fluid—sometimes oil, sometimes molten salt—flowing through a pipe to a central station, where the energy it transports can be converted to steam and used to drive a conventional turbine, similar to the strategy employed with the central tower option.

As of early 2014, globally installed CSP amounted to 3.65 GW with 1.17 GW in the United States, significantly less in both cases than the commitment to utility-scale PV. A further 0.64 GWs of CSP facilities are currently under construction in the United States. A summary of the major CSP plants deployed globally as of 2014 is presented in Table 11.2.

TABLE 11.2

A Summary of the Major Concentrated Solar Power Plants Deployed Globally as of 2014

Capacity (MW)	Name	Country	Location	Technology Type	Notations and References
392	Ivanpah Solar Power Facility	United States	San Bernardino County, California	Solar power tower	Completed in February 13, 2014
354	Solar Energy Generating Systems	United States	Mojave Desert, California	Parabolic trough	Collection of nine units
280	Solana Generating Station	United States	Gila Bend, Arizona	Parabolic trough	Completed in October 2013, with 6h thermal energy storage
200	Solaben Solar Power Station	Spain	Logrosán	Parabolic trough	Solaben 3 completed June 2012; Solaben 2 completed October 2012; Solaben 1 and 6 completed September 2013
150	Solnova Solar Power Station	Spain	Sanlúcar la Mayor	Parabolic trough	Solnova 1 completed May 2010; Solnova 3 completed May 2010; Solnova 4 completed August 2010
150	Andasol solar power station	Spain	Guadix	Parabolic trough	Andasol 1 completed, 2008, with 7.5 h thermal energy storage; Andasol 2 completed, 2009, with 7.5 h thermal energy storage; Andasol 3 completed, 2011, with 7.5 h thermal energy storage
150	Extresol Solar Power Station	Spain	Torre de Miguel Sesmero	Parabolic trough	Extresol 1 completed February 2010, with 7.5 h thermal energy storage; Extresol 2 completed December 2010, with 7.5 h thermal energy storage; Extresol 3 completed August 2012, with 7.5 h thermal energy storage

100	Palma del Rio Solar Power Station	Spain	Palma del Río	Parabolic trough	Palma del Rio 2 completed December 2010; Palma del Rio 1 completed July 2011
100	Manchasol Power Station	Spain	Alcázar de San Juan	Parabolic trough	Manchasol-1 completed January 2011, with 7.5 h heat storage; Manchasol-2 completed April 2011, with 7.5 h heat storage
100	Valle Solar Power Station	Spain	San José del Valle	Parabolic trough	Completed December 2011, with 7.5 h heat storage
100	Helioenergy Solar Power Station	Spain	Écija	Parabolic trough	Helioenergy 1 completed September 2011; Helioenergy 2 completed January 2012
100	Aste Solar Power Station	Spain	Alcázar de San Juan	Parabolic trough	Aste 1A completed January 2012, with 8 h heat storage; Aste 1B completed January 2012, with 8 h heat storage
100	Solacor Solar Power Station	Spain	El Carpio	Parabolic trough	Solacor 1 completed February 2012; Solacor 2 completed March 2012
100	Helios Solar Power Station	Spain	Puerto Lápice	Parabolic trough	Helios 1 completed May 2012; Helios 2 completed August 2012
100	Shams	United Arab Emirates	Abu Dhabi Madinat Zayad	Parabolic trough	Shams 1 completed March 2013
100	Termosol Solar Power Station	Spain	Navalvillar de Pela	Parabolic trough	Both Termosol 1 and 2 completed in 2013

Source: http://en.wikipedia.org/wiki/List_of_solar_thermal_power_stations.

The Ivanpah Solar Electric Generating System (ISEGS), which began operation in California's Mojave Desert in 2013, is presently the world's largest CSP plant with a rated capacity of 377 MW. A picture of the plants that compose this facility, which employ a tower design, is presented in Figure 11.5. Notable is the central tower that rises 459 feet above the desert floor. The Shams 1 plant in Abu Dhabi, which uses the parabolic trough option, is displayed in Figure 11.6. The pictures provide a striking indication of the massive scale of these projects. Each MW of CHP capacity requires a concentration of mirrors occupying an area of between 3 and 8 acres (Mendelson et al. 2012). The Ivanpah complex alone extends over an area of more than 3,500 acres—close to 5.5 square miles. The area required for utility-scale solar facilities is large, much greater than that required for comparable wind systems. It is probable, though, that they will be established in regions of limited economic value—deserts, for example—although there may be complex impacts on the environment that should be considered in the selection of suitable sites.

As indicated in Table 9.2, the levelized cost for electricity generated using CSP facilities projected to be operational in 2018 is significantly higher than the corresponding cost for PV: 26.2 cents per kWh as compared to 14.4 cents per kWh (2011 currency). There are important advantages, however, for CSP. First, energy harnessed from the sun during the day can be stored with CSP systems (in the form of molten salt, for example), in some cases overnight, allowing plants to produce electricity on a 24-hour schedule.

FIGURE 11.5 Plants that compose the Ivanpah Solar Electric Generating System (ISEGS). (*Source:* http://www.technocrazed.com/ivanpah-solar-power-plant-can-provide-electricity-to-140000-homes-and-roast-birds)

FIGURE 11.6 The Shams 1 concentrated solar energy power plant in Abu Dhabi features more than 258,000 mirrors mounted on 768 tracking parabolic trough collectors, covering an area of 2.5 km². (*Source:* http://humansarefree.com/2013/07/worlds-largest-solar-power-plant-opens.html)

Second, CSP plants can be designed from the outset to incorporate conventional gas-fired steam generators complementing production from sunlight, providing thus additional flexibility for the plant to respond to variations in demand for power. In contrast, electricity generated with PV, in the absence of facilities for storage of power (batteries, for example), must be dispatched in real time, significantly reducing the ability of utilities to respond in this case to changing patterns of demand.

KEY POINTS

1. The cost for PV systems required to convert solar radiation to electricity has declined steeply in recent years, primarily in response to overproduction in China.
2. While costs for acquisition of PV panels have fallen globally, to less than $1 per watt, overall costs for installation of these systems, including so-called soft costs, have remained stubbornly high, especially in the United States, where they average $5.20 per watt, almost twice the prevailing cost in Germany.
3. Third-party brokers, better equipped than individuals to take advantage of the financial incentives available to support investments in solar energy, have played an important role in facilitating the recent increase in household installations of PV systems in the United States.

4. There has been an important increase in deployment of PV systems by utilities taking advantage of both reductions in the price for components and economies of scale associated with deployment of these components in large installations.
5. Focusing sunlight using large arrays of mirrors provides a source of heat that can be deployed to generate steam to run conventional steam-fueled turbines. Electricity produced using this option, referred to as concentrated solar power (CSP), is more expensive than electricity generated using PV. CSP has the advantage that with this option energy can be stored in the form of heat, providing important flexibility for plant operators to delay production of electricity to respond more efficiently to changing patterns of customer demand.

Notes

1. Sharing of electrons is the key that allows atoms to combine to form more complex structures—molecules. Consider what happens if a silicon atom loses one of its electrons. It is transformed then into a positively charged product. If this positively charged atom is to link up with a positively charged neighbor, there needs to be a concentration of negative charge in the space in between to prevent the positively charged atoms from flying apart. The pair of shared electrons—one from each atom—is what provides the bond that connects the atoms.

2. Energy on an atomic scale is expressed conventionally in units of electron volts (eV), the energy an electron would acquire if exposed to an electrostatic potential of 1 volt (V): 1 eV is equal to 1.602×10^{-19} joule (J) (see Chapter 2 for definition of the joule). The shorter the wavelength of light, the greater its energy.

3. If an element tends to lose electrons when forming a compound, it is said to be oxidized. If it gains electrons, it is reduced. The carbon atom in CO_2 is oxidized: it has given up electrons to stabilize the links to the companion O atoms. The carbon atom in CH_4, on the other hand, is reduced: it has acquired electrons from the companion H atoms. The silicon atom in SiO_2 is oxidized: in the pure Si form it is reduced.

4. A heavy mass suspended from a tall crane possesses what is referred to as potential energy. If released, the mass would fall to the ground, picking up speed reflected in what is referred to as kinetic energy. It acquires this kinetic energy as a result of work done on the falling mass by gravity. Voltage is the electrostatic equivalent of potential energy. The greater the voltage difference between two positions on a conducting wire, the greater the energy generated by the electrically charged components of the wire. Voltage is a measure of the energy communicated to unit charge, joules per coulomb (JC^{-1}), where J defines the basic unit of energy in the SI system (cf. Chapter 2) and C is the corresponding SI unit for charge. Voltage in the SI system is expressed in volts indicated by the symbol V. The number of charged particles passing a particular position in a conducting wire is referred to as the current ($C\ s^{-1}$). The unit in this case is the amp, written as A. The product of current and voltage defines the power delivered by the current, energy per unit time. With energy expressed in joules and time in seconds, power is measured in watts (W).

5. PPAs represent contracts between producers of renewable energy and utilities that are obliged to include some fraction of renewable energy in the power they dispatch. These contracts,

which can extend over many years, guarantee a steady and reliable revenue stream to developers, lowering accordingly the risk profile for their investments. The incentives represented by PPAs have played an important role in recent years in the United States in stimulating development of renewable energy resources such as solar that might have been considered otherwise to have involved excessive risk.

References

Barbose, G., N. Darghouth, S. Weaver, and R. Wiser. 2013. *Tracking the sun VI: An historical summary of the installed price of photovoltaics in the United States from 1998 to 2012.* Berkeley, CA: Lawrence Berkeley National Laboratory.

Bloomberg New Energy Finance. 2013. Solar to add more megawatts than wind in 2013, for first time. http://about.bnef.com/press-releases/solar-to-add-more-megawatts-than-wind-in-2013-for-first-time/

Friedman, B., K. Ardani, D. Feldman, R. Citron, R. Margolis, and J. Zuboy. 2013. *Benchmarking non-hardware balance-of-system (soft) costs for U.S. photovoltaic systems, using a bottom-up approach and installer survey.* Golden, CO: National Renewable Energy Laboratory.

Masters, G. M. 2004. *Renewable and efficient electric power systems.* Hoboken, NJ: John Wiley & Sons, Inc.

McGinn, D., E. Macías Galán, D. Green, L. Junfeng, R. Hinrichs-Rahlwes, S. Sawyer, M. Sander et al. 2013. *Renewables 2013 global status report.* Paris: REN21.

Mendelson, M., T. Lowder, and B. Canavan. 2012. *Utility-scale concentrating solar power and photovoltaics projects: A technology and market overview.* Golden, CO: National Renewable Energy Laboratory.

Seel, J., G. Barbose, and R. Wiser. 2013. *Why are residential pv prices in Germany so much lower than in the United States? A scoping analysis.* Berkeley, CA: Lawrence Berkeley National Laboratory.

12

Hydro

POWER FROM RUNNING WATER

AS DISCUSSED IN Chapter 4 and illustrated in Figure 4.1, close to 50% of the solar energy intercepted by the Earth is absorbed at the surface. Approximately half of this energy, 78 W m^{-2}, is used to evaporate water, mainly from the ocean. What this means is that evaporation of water accounts for as much as a third of the total solar energy absorbed by the Earth (atmosphere plus surface). The atmosphere has a limited ability to retain this water. Evaporation is balanced in close to real time by precipitation. A portion of this precipitation reaches the surface in regions elevated with respect to sea level—in mountainous locations, for example. It is endowed in this case with what we refer to as potential energy (Chapter 4). This potential energy can be stored (in lakes or dams, for instance), or it can be released, converted to kinetic energy (directed motion) as the water flows downhill on its return to the ocean. And along the way, energy can be captured and channeled to perform useful work.

An early application involved exploiting the power of running water to turn a flat stone, one of two that constituted the apparatus used to grind grain, the other remaining stationary during the grinding process. The Domesday Book records that by AD 1086 as many as 5,624 water mills were operational in England south of the River Trent, deployed not just to grind grain but for a multitude of other tasks, including, but not confined to, sawing wood, crushing ore, and pumping the bellows of industrial furnaces (Derry and Williams 1960). Later, running water would provide the motive force for the textile industry that marked the beginning of the industrial age in North America, specifically in New England (Steinberg 1991; McElroy 2010). The most important contemporary application of water power involves the generation of electricity, the bulk of which is obtained by tapping the potential energy stored in high-altitude dams, a lesser fraction from the kinetic energy supplied by free-flowing streams (what is referred to as run-of-the-river sources).

The operational elements of a typical dam-based hydro system are illustrated in Figure 12.1. The dam serves to intercept the downstream transfer of water resulting from upstream precipitation. Water builds up behind the dam, often flooding the upstream environment and creating what amounts to an artificial lake. The quantity of potential energy stored by the dam depends on the elevation of the water above the penstock outlet. Water flows into the penstock with kinetic energy defined by the pressure at the outlet, determined in turn by the height of the overlying water. The kinetic energy of water in the penstock is deployed to turn the blades of a turbine with consequent production of electricity as indicated.

A picture of the Hoover Dam, formed by damming the Colorado River, is displayed in Figure 12.2. When completed in 1936, Hoover Dam was both the world's largest electricity-generating system and also the world's largest concrete structure. As a testament to the developments that occurred subsequently, the Hoover Dam now barely makes the list of the world's top 50 power-producing hydro developments (cf. Table 12.1 below). The picture provides an instructive illustration of the scale of the dam. Notable is the difference between the elevation of the water behind the dam and the level of water in the drainage outlet. When full, Lake Mead, the lake formed by the dam, would cover an area of 158,000 acres or 248 square miles. At its deepest point, the water would extend to a depth of more than 500 feet and the shoreline would range over more than 750 miles. The level of water in the Lake has dropped by more than 100 feet over the past

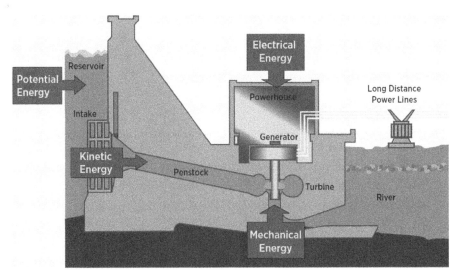

FIGURE 12.1 Operational elements of a typical dam-based hydro system.
(*Source:* http://www.alternative-energy-news.info/technology/hydro)

FIGURE 12.2 Picture of the Hoover Dam, formed by damming the Colorado River. (*Source:* http://commons.wikimedia.org/wiki/File:Hoovernewbridge.jpg)

12 years, reflecting the combination of a climate-related decrease in the supply of water from melting snow in the headwaters of the River and increased withdrawals to supply the growing demands for water in the neighboring states of Arizona, California, and Nevada.

The chapter begins with an account of the status of hydropower on a global scale followed by discussions of current conditions and future prospects for the United States and China, concluding with a summary of key points.

GLOBAL PERSPECTIVE

Hydropower accounted for 16.5% of the total electricity generated worldwide in 2011, 79% of all of the electricity produced from renewable sources (wind, 10.1%; biomass, 8.1%; geothermal, 1.5%; solar plus tidal, 1.3%). China led the way in terms of individual country production, responsible for 19.8% of the global total, followed by Brazil (12.2%), Canada (10.7%), and the United States (9.2%). On a global basis, as illustrated in Figure 12.3, production of electricity from hydro increased by more than 50% between 1990 and 2008, with China responsible for the largest fraction of this growth.

A list of the world's 50 largest hydro facilities is presented in Table 12.1 (htpp://en.wikipedia.org/wiki/List_of_largest_hydroelectric_power_stations). The Three Gorges

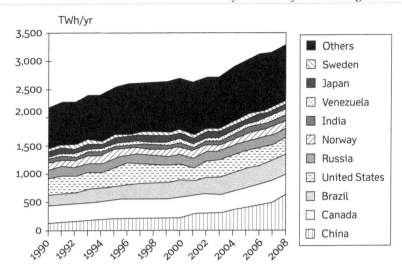

FIGURE 12.3 Global production of electricity from hydro from 1990 to 2008.

Dam on the Yangtze River in China ranks number 1 both in terms of capacity (22.5 GW) and average annual production of electricity (98.5 TWh). The only US facility to make the list of the top 10 in terms of capacity is the Grand Coulee Dam on the Columbia River in Washington State. The Grand Coulee, constructed over a 7-year period beginning on December 6, 1935, as part of President Roosevelt's New Deal, includes four powerhouses incorporating a total of 33 individual generators. The original facility had a capacity of 1.974 GW. It was expanded subsequently to 6.809 GW. In terms of power generation, Grand Coulee, with an annual average power production of 20 TWh, currently ranks number 11 on the list of the major global hydro facilities included in Table 12.1. The Itaipu Dam on the border between Brazil and Paraguay ranks number 2, trailing only the Three Gorges facility, with an annual average output of 98.3 TWh.

Production of electricity represents but one of a number of the functions served by dams on major rivers. This multifunction motivation was recognized in the congressional legislation in the United States that authorized (on August 20, 1938) construction of the Coulee and Parker Dams, the latter located on the Colorado River 155 miles downstream from the Hoover Dam (HR 625). The wording of the resolution was as follows:

That for the purpose of controlling floods, improving navigation, regulating the flow of the streams of the United States, providing for storage and for the delivery of the stored waters thereof, for the reclamation of public lands and Indian reservations, and other beneficial uses, and for the generation of electric energy as a means of financially aiding and assisting such undertakings the projects known as "Parker Dam" on the Colorado River and "Grand Coulee Dam" on the Columbia River are hereby authorized and adopted.

TABLE 12.1

World's 50 Largest Hydropower Facilities

Rank	Name and Years of Completion	Country and River	Installed Capacity (MW)	Annual Electricity Production (TWh)	Area Flooded (km²)
1	Three Gorges Dam 2003/2012	P.R. China Yangtze	22,500	98.5	632
2	Itaipu Dam 1984/1991, 2003	Brazil Paraguay Paraná	14,000	98.3	1,350
3	Guri 1978, 1986	Venezuela Caroní	8,850	53.41	4,250
4	Tucuruí 1984	Brazil Tocantins	8,370	41.43	3,014
5	Grand Coulee 1942/1950, 1973, 1975/1980, 1984/1985	United States Columbia	6,809	20	324
6	Longtan Dam 2007/2009	P.R. China Hongshui	6,426	18.7	
7	Krasnoyarsk 1972	Russia Yenisei	6,000	20.4	2,000
8	Robert-Bourassa 1979/1981	Canada La Grande	5,616	26.5	2,835
9	Churchill Falls 1971/1974	Canada Churchill	5,428	35	6,988
10	Bratsk 1967	Russia Angara	4,500	22.6	5,470
11	Laxiwa Dam 2010	P.R. China Yellow	4,200	10.2	

12	Xiaowan Dam
2010	P.R. China
Mekong	4,200	19	190
13	Sayano–Shushenskaya
1985/1989, 2010/2014	Russia
Yenisei	3,840	26.8	621
14	Ust Ilimskaya
1980	Russia
Angara	3,840	21.7	
15	Tarbela Dam
1976	Pakistan
Indus	3,478	13	250
16	Ilha Solteira Dam
1973	Brazil
Paraná	3,444	17.9	
17	Ertan Dam
1999	P.R. China
Yalong	3,300	17	
17	Pubugou Dam
2009/2010	P.R. China
Dadu	3,300	14.6	
19	Macagua
1961, 1996	Venezuela
Caroní	3,167.5	15.2	47
20	Xingó Dam
1994/1997	Brazil
São Francisco	3,162	18.7	
21	Yacyretá
1994/1998, 2011	Argentina Paraguay
Paraná	3,100	20.09	1,600
22	Nurek Dam
1972/1979, 1988	Tajikistan
Vakhsh	3,015	11.2	98
23	Bath County PSP
1985, 2004	United States
—	3,003	3.32	

(*continued*)

TABLE 12.1
Continued

Rank	Name and Years of Completion	Country and River	Installed Capacity (MW)	Annual Electricity Production (TWh)	Area Flooded (km²)
24	Goupitan Dam 2009/2011	P.R. China Wu	3,000	9.67	94
25	W. A. C. Bennett Dam 1968, 2012	Canada Peace River	2,876	13.1	
26	La Grande-4 1986	Canada La Grande	2,779		765
27	Gezhouba Dam 1988	P.R. China Yangtze	2,715	17.01	
28	Manic-5 and Manic-5-PA 1970/1971, 1989/1990	Canada Manicouagan	2,656		1,950
29	Chief Joseph Dam 1958/73/79	United States Columbia	2,620	12.5	34
30	Volzhskaya (Volgogradskaya) 1961	Russia Volga	2,582.5	10.43	
31	Niagara Falls (United States) 1961	United States Niagara	2,525		
32	Revelstoke Dam 1984, 2011	Canada Columbia	2,480		115
33	Paulo Afonso IV 1979/1983	Brazil São Francisco	2,462.4		

34	Chicoasén (Manuel M. Torres) Dam 1980, 2005	Mexico	Grijalva	2,430	
35	La Grande-3 1984	Canada	La Grande	2,418	
36	Atatürk Dam 1990	Turkey	Euphrates	2,400	8.9
36	Jinanqiao Dam 2010	P.R. China	Jinsha	2,400	
36	Sơn La Dam 2010/2012	Vietnam	Black	2,400	10.25
36	Bakun Dam 2011	Malaysia	Balui	2,400	
36	Liyuan Dam 2012	P.R. China	Jinsha	2,400	
36	Guandi Dam 2013	P.R. China	Yalong	2,400	
42	Zhiguliovskaya (Samarskaya) 1957	Russia	Volga	2,335	8.8
44	Karun III Dam 2005	Iran	Karun	2,280	4.17
44	Iron Gates-I 1970	Romania Serbia	Danube	2,192	11.3
45	Caruachi 2006	Venezuela	Caroní	2,160	12.95

(*continued*)

TABLE 12.1
Continued

Rank	Name and Years of Completion	Country and River	Installed Capacity (MW)	Annual Electricity Production (TWh)	Area Flooded (km^2)
45	John Day Dam 1949	United States Columbia	2,160		
47	La Grande-2-A 1992	Canada La Grande	2,106		
48	Aswan 1970	Egypt Nile	2,100	11	
49	Itumbiara 1980	Brazil Paranaíba	2,082		
50	Hoover Dam 1936/1939, 1961	United States Colorado River	2,080	4	

Source: http://en.wikipedia.org/wiki/List_of_largest_hydroelectric_power_stations.

Production of electricity was recognized at least at that time as an ancillary function of the dams, one that could serve to compensate in part for the expense of their construction.

UNITED STATES: CURRENT STATUS AND PROSPECTS

There are approximately 80,000 dams on rivers in the United States. Only a small fraction, however, about 3%, has been tapped to produce electricity. The current capacity of hydropower facilities in the United States amounts to about 78 GW with an additional 22 GW available in the form of pumped hydro. Hydropower is responsible for approximately 7% of the total electricity consumed in the United States. Production varies both seasonally and interannually, responding to a combination of climate and weather-related variability in the supply of water.

Table 12.2 presents a list of the 10 largest hydro facilities in the United States (http://ussdams.org/uscold_s.html). More than half of the total US capacity for production of electricity from hydropower is concentrated in three western states—Washington, California, and Oregon. Washington alone is responsible for approximately 26% of the national total. Construction of large dams peaked in the United States in the 1960s, reflecting growing public awareness of the potential adverse effects of such projects on the environment. The opposition to the Glen Canyon Dam, located on the Colorado River upstream of the Grand Canyon, one of the most recently completed projects (1964), provides a case in point.

The objections are focused for the most part in this case on the impact of the dam on the ecology of the region downstream. Under natural conditions, the flow of water

TABLE 12.2

The 10 Largest Hydro Facilities in the United States

Dam Name	River	Location	Capacity (MW)
Grand Coulee	Columbia	Washington	6,180
Chief Joseph	Columbia	Washington	2,457
John Day	Columbia	Oregon	2,160
Bath County P/S	Little Back Creek	Virginia	2,100
Robert Moses—Niagara	Niagara	New York	1,950
The Dalles	Columbia	Oregon	1,805
Ludington	Lake Michigan	Michigan	1,872
Raccoon Mountain	Tennessee River	Tennessee	1,530
Hoover	Colorado	Nevada	1,434
Pyramid	California Aqueduct	California	1,250

Source: http://en.wikipedia.org/wiki/Hydroelectric_power_in_the_United_States.

downstream of the dam would be seasonally variable, punctuated by episodic major floods. With construction of the dam and the filling in of the major artificial lake behind it (Lake Powell), these floods have been largely eliminated. To offset the ecological disturbance occasioned by the dam, environmental groups have been calling in some cases for removal of the dam, in others, more moderately, for a management system that would regulate release of water from the dam to mimic as far as possible downstream preexisting (natural) conditions. Elsewhere there have been calls for elimination of four dams on the Snake River in Washington. The issue in this case relates to the negative impact these dams are perceived to have on the migration of fish, specifically salmon. Given the current political climate, prospects for construction of major new, dam-based, hydroelectric facilities in the United States would appear to be limited.

Kao et al. (2014), in a report commissioned for the Water Power Program of the US Department of Energy, estimated the capacity for untapped hydropower in the United States at 84.7 GW, with potential to supply on an annual average basis as much as 460 TWh of electricity. Excluding areas protected by Federal Law (National Parks and designated Wilderness Areas, for example), their estimate for the potential capacity of additional US hydropower drops to 65.5 GW, slightly lower than the current installed capacity of 79.5 GW, with the possibility to more than double the present source of US hydro-generated electricity (347 TWh per year as compared to 272 TWh per year). The study, as the authors point out, was based on the theoretical, physical potential of some 3 million of the nation's streams. It was not intended to determine the economic feasibility nor the public acceptance of any specific site for development. In an earlier report, the Department of Energy (http://energy.gov/articles/energy-department-report-finds-major-potential-increase-clean-hydroelectric-power) concluded that adding power-generating facilities to a fraction of the country's dams that currently lack this capability could increase the nation's hydropower capacity by as much as 15%, a prospect worthy of further consideration.

STATUS AND PROSPECTS FOR HYDROPOWER IN CHINA

As indicated, China leads the world in terms of both existing hydropower capacity and plans for future expansion. Development of its hydropower potential is a key element in China's plans to furnish as much as 15% of its primary energy demand from renewable sources by 2020. With completion of the Three Gorges Dam in 2012, the capacity of hydropower in China reached a record level of 249 GW. The 12th five-year plan (2011–2015) called for an increase to 325 GW by the end of 2015 and to 348 GW by 2020. The theoretical maximum hydropower potential of China is estimated at 694 GW, of which 402 GW is considered to be technically and economically exploitable, capable of producing 1,750 TWh of electricity on an annual basis (Huang and Yan 2009). To place this number in context, China consumed 4,819 TWh of electricity from all sources in 2012. EIA (2013) projects that demand will increase to 7,295.5 TWh by 2020, requiring an expansion of the overall generating capacity to 1588.6 GW as compared to the level of 1144 GW that was available in 2012, much of this expected to be supplied by conventional thermal sources.

The Three Gorges plant accounted for approximately 8% of the total electricity produced from hydropower in China in 2013. The Three Gorges Dam stands close to 600 feet in height, 1.3 miles in width. Some 27.2 million cubic meters of concrete and 463,000 tons of steel were deployed in its construction. The reservoir behind the dam covers an area of more than 403 square miles, raising water levels significantly to as far upstream as Chongqing, 360 miles above the dam. Some 1.3 million people were displaced in the course of the dam's construction. More than 100 towns were submerged. Construction of the dam was controversial both within and outside of China. Objections focused on the environmental and societal disruptions that would accompany both the construction and subsequent operation of the dam. And there were concerns that the dam could pose a threat to the security of communities downstream in the event that it could be targeted for demolition by terrorists, by future enemies of the state, or by an unanticipated disaster of natural origin (an earthquake, for example, either locally or upstream causing rupture of upstream dams). Significantly—and unusually—this opposition was expressed at the meeting of the National People's Congress that finally approved the project in 1992. Of the 2,633 delegates authorized to vote at that meeting, only 1,767 elected to register their votes in favor.

The Three Gorges Dam was proposed in its design phase to address multiple objectives. The 34 generators it incorporates can accommodate a power output as great as 22.5 GW. In practice, the power source is variable, responding to the seasonal pattern of flows in the river feeding the dam—a maximum in summer, a minimum in winter. The dam incorporates a series of locks that make it easier for ships of significant size to navigate all the way from Shanghai to Chongqing. The existence of the extensive lake behind the dam increases the safety of transit through the Gorges for these ships, an often-hazardous passage prior to construction of the dam. The dam contributes also in a significant way to decreasing the risk of flooding in downstream areas while at the same time providing a source of water to compensate for periods of drought.

Future plans for development of China's hydro resources involve construction of a series of dams on the Jinsha River, the upper reach of the Yangtze River, anticipated to add as much as 75 GW to the country's present hydropower capacity. Plans on the drawing board call for additional projects to exploit the potential resources of the Mekong River (extending from China through Myanmar, Laos, Thailand, Cambodia, and Vietnam), the Nu River (known as the Salween River in Myanmar, extending from China through Myanmar, forming on a portion of its transit the border between Myanmar and Thailand), and the Brahmaputra River (extending from China through India, joining up with the Ganges River in Bangladesh). Development of these resources could add as much as a further 50 to 75 GW to China's ultimate source of hydropower. All of these rivers originate in the Tibetan Plateau. Runoff is seasonally variable. Production of electricity varies accordingly, a pattern reflected in the production of electricity from the country's existing hydropower infrastructure, as indicated in Figure 12.4.

Production of electricity from hydropower in China peaks typically in summer with a minimum during the driest months in winter. As indicated in the figure, the amplitude of the seasonal signal has increased significantly over the past 25 years, from approximately ±30% of the annual average in the early 1980s to ±50% in the more recent decades. This trend reflects primarily climate-related variations in the supply of waters to the country's hydro facilities. The problem was particularly severe during the summer of 2011, as indicated in Figure 12.5. China suffered through a drought in that year, the worst on record since the 1950s. Production of electricity from hydropower contracted by between 16% and 25% with respect to the prior norm, resulting in a 4% to 5% decrease in China's total national production of electricity, a shortfall that could be made up only by engaging as much as an additional 39 GW of conventional (mainly thermal) power sources operating at an average capacity factor of 50% (Deutsche Bank Group 2011).

The experience in 2011 provides a dramatic illustration of the interconnection between climate (notably variability in the relatively localized regions of the Tibetan Plateau that provide the dominant feed for the relevant major rivers) and the potential contribution of hydro to China's future demand for electricity. Engineers charged with designing and sizing dams scheduled for future construction are obliged to rely on historical data for the flows of waters in relevant rivers. But, in the face of likely changes in future climate, this information could be misleading. Dams constructed on the basis of this input could turn out to be either too small or too large to take optimal advantage of future water

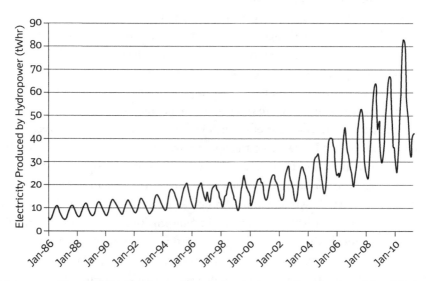

FIGURE 12.4 Temporal trends and seasonal variations in the overall production of electricity from hydropower in China (1986–2010).
(*Source:* Deutsche Bank Group 2011)

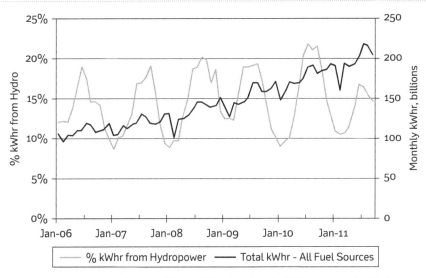

FIGURE 12.5 Increase in total national demand for electricity in China compared with the fractional supply from hydro.
(*Source:* Deutsche Bank Group 2011)

supplies. In the former case, exposure to floods of unanticipated magnitude could place the structural integrity of the dams at risk. In the latter case, the return on investment for construction of the dams could be less than required to justify the original expense.

KEY POINTS

1. China leads the world in terms of production of electricity from hydropower, followed by Brazil and Canada, with the United States in fourth place.
2. Approximately 7% of electricity consumed in the United States is produced from hydro sources, as compared to 17% for China.
3. Prospects for future expansion of hydropower in the United States are limited. With concerns as to their environmental impact, there is little support for construction of major new dams.
4. There are approximately 80,000 small dams on rivers in the United States, only a fraction of which, about 3%, have been tapped to produce electricity. It is estimated that instrumenting a small fraction of these dams to produce electricity could enhance production of electricity from hydropower in the United States by as much as 15%.
5. China has ambitious plans to increase its production of electricity from hydropower, adding as much as 75 GW to its current installed capacity of 250 GW by 2015 and 100 GW by 2020. Developing its hydropower resources, cutting back on the demand for fossil fuel–based power sources of electricity, and thus curtailing associated emissions of climate-altering CO_2 is an important

element in China's plan to supply up to 15% of its primary energy demand from renewable sources by 2020.
6. Production of electricity from hydro resources is seasonally variable both in China and in the United States, sensitive in both cases to potential changes in climate impacting supplies of water to both present and prospective future hydro systems.
7. In addition to their role as a source of electric power, dams provide a variety of ancillary functions, including their contribution in supplying water to neighboring communities, controlling floods, improving navigation, and serving as a source of enhanced recreational facilities.
8. Offsetting these advantages are costs associated with land lost during construction of the dams, penalties for populations that have to be relocated, impacts on fish and wildlife, and consequences for downstream ecological communities resulting from changes in the seasonal pattern of flows of water in impacted rivers.

References

Derry, T. K., and Williams, T. I. 1960. *A short history of technology: From the earliest times to A.D. 1900*. Oxford: Oxford University Press.

Deutsche Bank Group. 2011. *Hydropower in China: Opportunities and risks*. New York: Deutsche Bank Group.

Kao, S.-C., R. A. McManamay, K. M. Stewart, N. M. Samu, B. Hadjerioua, S. T. DeNeale, D. Yeasmin, M. Fayzul, K. Pasha, A. A. Oubeidillah, and B. T. Smith. 2014. *New stream-reach development: A comprehensive assessment of hydropower energy potential in the United States*. Oak Ridge, TN: Oak Ridge National Laboratory.

McElroy, M. B. 2010. *Energy perspectives, problems, and prospects*. New York: Oxford University Press.

Steinberg, T. 1991. *Nature incorporated: Industrialization and the waters of New England*. Cambridge: Cambridge University Press.

13

Earth Heat and Lunar Gravity

GEOTHERMAL AND TIDAL ENERGY

TO THIS POINT, we have discussed the current status and future prospects of energy from coal, oil, natural gas, nuclear, wind, solar, and hydro. With the exception of the contribution from nuclear, the ultimate origin of the energy for all of these sources is the sun—energy captured millions of years ago by photosynthesis in the case of the fossil fuels (coal, oil, and natural gas), energy harvested from contemporary inputs in the case of wind and solar. We turn now to a discussion of the potential for generation of electricity from geothermal sources and ocean tides. Decay of radioactive elements in the Earth's interior provides the dominant source for the former; energy extracted from the gravitational interaction of the Earth and moon is the primary source for the latter.

There are two main contributions to the energy reaching the surface from the Earth's interior. The first involves convection and conduction of heat from the mantle and core. The second reflects the contribution from decay of radioactive elements in the crust, notably uranium, thorium, and potassium. The composite geothermal source, averaged over the Earth, amounts to about 8×10^{-2} W m^{-2}, approximately 3,000 times less than the energy absorbed from the sun. As a consequence of the presence of the internal source, temperatures increase at an average rate of about 25°C per kilometer as a function of depth below the Earth's surface. The rate of increase is greater in regions that are tectonically active, notably in the western United States and in the region surrounding the Pacific Ocean (the so-called Ring of Fire)—less in others. Of particular interest in terms of harvesting the internal energy source to produce electricity are hydrothermal reservoirs, subsurface environments characterized by the presence of significant quantities of high-temperature water formed by exposure to lava or through contact with unusually hot crustal material.

The water contained in hydrothermal reservoirs is supplied for the most part by percolation from the surface through overlying porous rock. The conditions required for production of these hydrothermal systems are relatively specialized. The porosity of the rock must be sufficient to allow the water to penetrate to a depth sufficient to ensure contact with high-temperature rock or lava. This is likely to occur only in regions that are either presently or recently tectonically active. If the rock conditions are favorable, the superheated water may be returned to the surface in the form of hot springs, geysers, or fumaroles, primarily liquid in the first two cases, liquid converted to vapor before it reaches the surface in the third. Alternatively, it may be trapped in an impermeable rock formation overlying the zone where the hot water is produced in the first place.

There is a long history to the use of hot springs as a source of heated water for bathing, dating back to the Qin Dynasty in China in the third century BC. The Romans used hot springs in Bath in England in the first century AD to supply hot water to public baths and as a source of hot water for under-floor heating of buildings. Geothermal energy has been deployed as a source of district heating in Chaudes-Aigues, France, for more than 700 years, and it has been exploited more recently, in 1892, for a similar purpose in Boise, Idaho, in the United States. Geothermal energy was first used to produce electricity in Larderello, Italy, in 1911. The source in this case was steam from a regional fumarole field. An instructive summary of the history of geothermal use is presented online (https://en.wikipedia.org/wiki/Geothermal_energy).

There are a number of ways in which geothermal energy can be harnessed to produce electricity. If the natural product of a hydrothermal system is steam, the steam can be deployed directly to drive a conventional turbine. The plant in this case is referred to as a dry steam installation. If the product is hot water, steam can be generated by lowering the pressure on the water. Again, steam can provide the motive force for the turbine. If the temperature of the water is too low for this flashing process to produce steam, the water can be passed through a heat exchanger and used to heat a liquid such as isobutane with a lower boiling point than water. Vapor from the secondary liquid can be used then to drive the turbine. This option defines what is referred to as a binary approach. In practice, all three strategies are used in different installations depending on the nature of the geothermal resource. If steam or hot water from the reservoir is likely to be depleted, it can be supplemented by injecting water into the source reservoir.

The largest power-generating geothermal facility in the United States is located north of San Francisco in a region known as The Geysers. The natural product of this system, despite its name, is steam. The source is a magma chamber located more than 4 miles below the surface, measuring more than 8 miles in diameter. The first well for power generation was developed there in 1924. Deeper wells were drilled in the 1950s, but the major development took place in the 1970s and 1980s in response to the rapid rise in prices for oil triggered by the Arab boycott in 1973 and later in the decade by the instability prompted by the fall of the Shah and the subsequent hostage crisis in Iran (as discussed in Chapter 7). The current generating capacity at The Geysers amounts to about 2 GW. Treated wastewater from surrounding communities is injected into the reservoir in order to maintain a continuing, economically productive supply of steam.

Water and steam produced from hydrothermal sources include dissolved salts and a variety of toxic elements such as boron, lead, and arsenic, in addition to gases such as hydrogen sulfide and carbon dioxide. Care must be exercised to limit release of these chemicals to the environment. The problem is less serious in the case of binary systems where the water brought to the surface is readily isolated and can be returned to the hydrothermal source without release of constituent, potentially offensive pollutants.

As indicated, the gravitational interaction between the Earth and the moon, with an assist from the interaction of the Earth with the sun, is the primary motive force for the rise and fall of ocean water associated with global tides. Typically there are two tidal oscillations per day—two highs and two lows. One of the high water excursions is located on the side of the Earth closest to the moon. Water, which is more mobile than the solid earth, responds differentially to the lunar force, resulting in an accumulation of water and consequently a generally high tide in the sublunar region. It is more difficult to account for the second high tide that develops on the side of the Earth opposite from the moon. The explanation in this case is that the pull of the moon on the solid earth tends to leave water stranded: the result, a second high tide. The tidal highs are compensated by tidal lows elsewhere, accounting for the characteristic (approximate) 12-hour period associated with the ocean tides.

As the water sloshes back and forth in response to the gravitational pull of the moon, energy is dissipated, mainly through friction with the underlying sea bottom and the landforms with which the water comes into contact. This loss of energy is reflected in a slow change in the rotation of the Earth together with an increase in the mean distance separating the Earth and the moon. The length of the day increases as a result by approximately 1 second every 40,000 years. During the Carboniferous period 350 million years ago, the day-night cycle took approximately 23 hours rather than the contemporary 24 hours: the year was correspondingly longer, 385 days as compared with the present standard of 365.25 days (with shorter days it took more days for the Earth to complete its orbit around the sun). On a human, if not on a geological time scale, these changes are relatively inconsequential.

The tidal amplitude varies from place to place, influenced by the configuration of the seafloor and the shape of the continental landmasses with which the tidal water comes into contact. It is particularly high in embayments with narrow openings to the ocean. The amplitude of tides experienced in the Bay of Fundy in Nova Scotia, Canada, ranges as high as 16 meters, more than 50 feet, responding to an unusual combination of factors operating in that environment (http://www.amusingplanet.com/2012/03/tides-at-bay-of-fundy.html).

196　Energy and Climate

The chapter begins with a global perspective on the current exploitation of geothermal energy for purposes of electric power production, continuing with a more specific account of the status of present development in the United States. This is followed with an account of the opportunities for significant expansion of the geothermal option in the future, making use of what are referred to as enhanced geothermal systems (EGSs). It continues with a discussion of the more limited potential for exploitation of tidal energy, concluding with a summary of the key points of the presentation.

GEOTHERMAL ENERGY: GLOBAL PERSPECTIVE

An instructive review by the Geothermal Energy Association (GEA) of the current state of the global geothermal industry is presented online (http://geo-energy.org/events/2014%20Annual%20US%20&%20Global%20Geothermal%20Power%20Production%20Report%20Final.pdf, read July 28, 2014). As of January 2014, the nameplate capacity for globally installed geothermal systems had grown to a record level of 12.013 GW. If plants under construction are completed on schedule, the report concludes that the total global capacity should grow to 13.45 GW by 2017. The International Energy Agency (IEA) (http://www.iea.org/files/ann_rep_sec/geo2010.pdf, read July 29, 2014) estimates the technical potential for production of electricity globally from tectonically active regions at 650 GW, assuming an efficiency for conversion of thermal energy to electricity of 10%. Resources that could be developed readily using conventional technology (sources with water temperatures greater than 130°C), assuming again a conversion efficiency of 10%, are estimated at 200 GW. These numbers may be compared with the current installed global capacity for production of electricity from all sources (fossil, nuclear, and renewable) of 5.3 TW.

The history of the development of global geothermal industry dating back to 2000 is displayed in Figure 13.1. The United States, with 3.44 GW of installed capacity, equivalent to 28.6% of the global total, ranks number 1. The situation for the top eight countries, including the United States, is summarized in Figure 13.2. Notable is the magnitude of the investments in the Philippines, Indonesia, and Mexico, in addition to Italy, New Zealand, Iceland, and Japan. All of these countries are located in regions that are tectonically active, benefitting accordingly from abundant geothermal resources. The growth rate for investments in geothermal power internationally is currently significantly greater than that in the United States. The leader in terms of plants under construction is Indonesia with 425 MW, followed by Kenya (296 MW), Iceland (260 MW), the United States (178 MW), New Zealand (166 MW), and the Philippines (110 MW). The GEA report suggests that given current trends, the installed capacities of Indonesia and potentially also the Philippines could be comparable to that of the United States in less than a few decades.

Geothermal energy has not to date played a significant role in China's plans to develop its electric power system. The environment, however, is changing. The central government has charged local authorities in northern, central, and southwestern regions of the

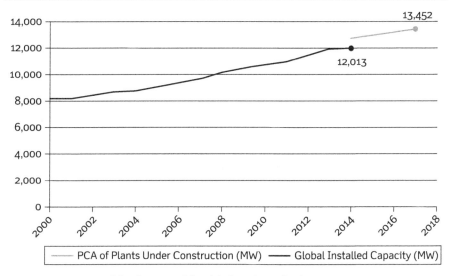

FIGURE 13.1 History of development of the global geothermal industry. (*Source:* GEA 2014)

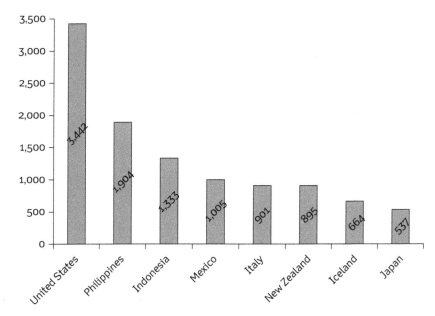

FIGURE 13.2 Top eight countries in terms of installed geothermal power in MW. (*Source:* GEA 2014)

country to draft plans for future development of the resource, suggesting a target of 100 MW for 2015 with additional expansion by 2017 (http://www.bloomberg.com/news/2014-07-10/china-calls-for-local-planning-on-geothermal-energy-use.html).

A breakdown of geothermal power production in terms of the technology deployed is presented in Figure 13.3 (geo-energy.org/events/2013%20International%20Report%20Final.pdf). Single-flash and dry-steam options account for more than half of total

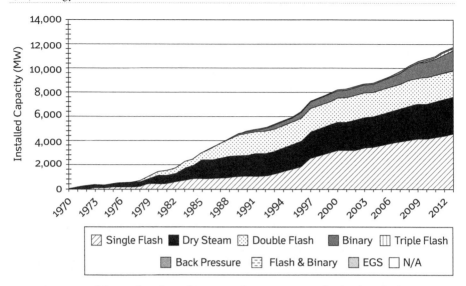

FIGURE 13.3 Breakdown of geothermal power production in terms of technology deployed. (*Source:* GEA 2013a)

current production. The single-flash approach involves lowering the pressure on the hot geothermal water source to produce steam, as discussed earlier. The double-flash option allows for production of a second source of steam by processing the water that remains after the first flash. The importance of the binary option (i.e., heating a secondary liquid), while continuing to implement single-flash, dry steam and double-flash deployments, is evidently increasing with time, reflecting presumably the opportunity to produce power from source waters at lower temperature.

GEOTHERMAL ENERGY: STATUS AND PROSPECTS FOR THE UNITED STATES

Not surprisingly, geothermal plants in the United States are concentrated mainly in the tectonically active west. As of February 2013, the total installed geothermal capacity amounted to 3.386 GW, a small fraction—about one third of a percent—of the nation's total power-generating capacity, but more significant for some of the western states, notably for California and Nevada. A breakdown of installed capacities by state is presented in Figure 13.4 (GEA 2013). California leads the way with 2.732 GW, followed by Nevada with 515 MW. Projects underway or at an advanced state of planning are expected to add approximately 1 GW to California's total geothermal capacity over the next few years. Development is even more rapid in Nevada, where the total installed geothermal capacity is projected to double over the next 3 years (GEA 2013).

A view of the choice of technology options for US geothermal power is presented in Figure 13.5 (GEA 2013). Dry steam, with 1,585 MW, dominates the present system, followed by flash and binary installations responsible for 997.3 MW and 803.57 MW, respectively. As indicated, as is the case globally, growth is most rapid for plants employing the binary option.

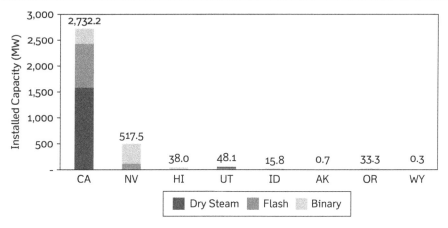

FIGURE 13.4 Breakdown of installed capacities by state in the United States. (*Source:* GEA 2013 b)

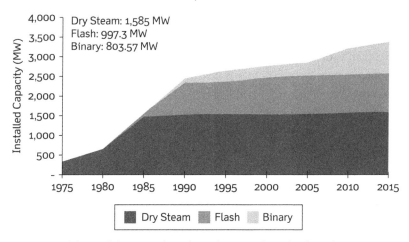

FIGURE 13.5 Breakdown of the national geothermal capacity by technology choice. (*Source:* GEA 2013b)

ENHANCED GEOTHERMAL SYSTEMS

Current geothermal plants operate for the most part by tapping into natural hydrothermal reservoirs. These systems account, however, for only a small fraction of the total potentially available source of geothermal energy: the EGS option would not be restricted to hydrothermal systems. The first step with EGS would be to identify environments that could be accessed at reasonable depth with temperatures high enough to provide the source of energy (heat) needed to produce the steam or hot water required for economically viable production of electricity, typically environments with temperatures greater than 150°C, preferably greater than 200°C. The water that could harness this energy would be supplied by drilling into the environment from the surface with a return flow through a separate conduit, thus eliminating the need to rely on a naturally occurring hydrothermal source. Water introduced in this manner would be heated by contact with the hot rock. For heat transfer to be effective, it would be important to maximize the surface area of the rock to which the water is exposed. To accomplish this objective, the water must be free to percolate through an extended network of open fractures. If such a network is not naturally present, it must be created artificially by hydraulically cracking the rock. A study by the Idaho National Laboratory on the future of EGS (INL 2006) suggests that to meet the energy demand of a 100 MW electricity plant at the surface, the volume of rock that would need to be altered at depth could range as high as 5 km^3.

The technical limit for current drilling technology extends to a depth of 10 km or even deeper (INL 2006). The distribution of temperature as a function of position across the continental United States is illustrated for depths of 6.5 km and 10 km in Figures 13.6 and 13.7. As expected and as indicated, conditions are most favorable in western regions of the country. INL (2006) suggests that early development of EGS in the United States would likely be concentrated in this region, probably in close proximity to existing hydrothermal developments to take advantage both of favorable environmental conditions and ease of access to existing transmission infrastructure. They suggest further that hot water developed in conjunction with conventional oil and gas production should be exploited as an additional EGS opportunity for production of electricity. Their study concludes that, with a relatively modest investment in research over the next decade or so, EGS could supply as much as 10% of demand for electricity projected for the United States in 2050. They estimate that the levelized cost for power generated with future EGS systems could drop in this case to as low as 4 and 6 cents per kWh, ensuring that EGS could compete on a cost basis with potential alternative sources of electricity.

While water is likely to provide the medium of choice for heat transfer for the majority of future geothermal applications, Brown (2000) and Pruess and Azaroual (2006)

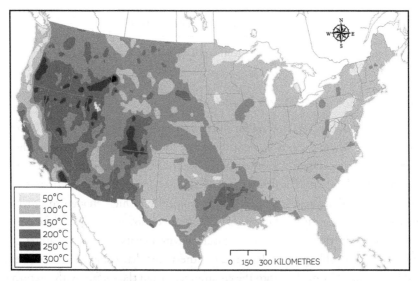

FIGURE 13.6 Distribution of temperature as a function of position across the continental United States at a depth of 6.5 km.
(*Source:* INL 2006)

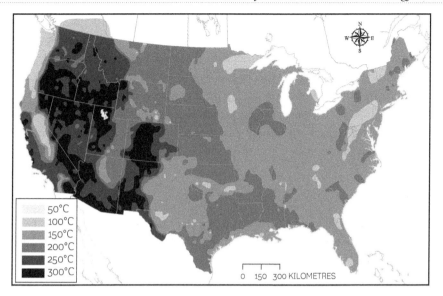

FIGURE 13.7 Distribution of temperature as a function of position across the continental United States at a depth of 10 km.
(*Source:* INL 2006)

outlined advantages that could be realized by using CO_2. The idea is that pressurized CO_2 could be introduced in liquid form into the fluid circulating system. As the temperature of this liquid is increased through contact with the hot rock at depth, its CO_2 composition would transition to a supercritical state defined by properties intermediate between those of a gas and a liquid. The transition to the supercritical state for CO_2 occurs at pressures higher than 72.9 atmospheres, at temperatures higher than 31.25°C, and there are reasons to believe that the heat transfer properties of supercritical CO_2 may be superior to those of water. The pressure would drop as hot CO_2 continues its circulatory path back to the surface. With the drop in pressure, the CO_2 content would make a second transition, in this case to the gas phase, in which form it could be deployed to drive a turbine and thus produce electricity. With a portion of the energy contained in the hot vapor expended in producing this electricity, the flow of CO_2 could be pressurized subsequently and converted to the liquid state and returned then to the hot rock at depth in an effectively closed circulatory loop. CO_2 could provide a useful substitute for water as a heat transfer agent under conditions where supplies of the latter are limited. The CO_2 deployed in this application could be produced by capture from smoke stacks of neighboring coal- or gas-fired power plants, providing thus a useful application for a product that would otherwise be released to the atmosphere with consequent implications for change in the global climate system.

China has significant resources for geothermal energy concentrated primarily in the country's west and southwest. As yet there is no agreed-upon national plan to exploit this potential for electric power production. The situation could change in the future, depending on experience gained from exploitation of the resource in the United States and elsewhere.

If the optimistic projections of INL (2006) for the future of geothermal power are in fact realized, EGS could make an important contribution to the future operation of the US electric power system. It could provide a source of baseload power that could cut back on the demand for coal, which, as discussed earlier, in combination with nuclear, plays the primary role in serving this function today. Furthermore, production from EGS systems could be programmed to compensate for the intrinsic variability of outputs from alternative renewable sources such as wind and solar. The advantages are significant. A persuasive case can be made for the modest investments in research recommended by INL (2006).

POWER FROM TIDES

Two strategies are employed to extract energy from tides to produce electricity. One makes direct use of the kinetic energy available in a fast flow of tidal water entering an embayment with a narrow opening to the ocean. The mechanism in this case is similar to that employed to produce electricity from wind. The second relies on a dam to build up a head of water in a basin behind the dam during the flood stage of the tidal cycle, with the potential energy developed in this manner deployed subsequently to produce electricity during the ebb phase as the water flows back to the ocean. The mechanism in this case is similar to that employed to generate electricity in a conventional river-based hydro system (see Chapter 12).

The 1.2 MW power-generating system at the mouth of Strangford Lough in Northern Ireland provides an example of the former. The speed of the water entering and leaving the Lough ranges as high as 4 meters per second (close to 10 miles per hour). The kinetic energy associated with this tidal motion is harnessed to turn the blades of an underwater turbine with a power rating of 1.2 MW. The direction of rotation of the blades is reversible: power is generated thus by harnessing kinetic energy on both the flood and ebb phases of the tide.

The facility on the Rance River in Brittany, France, provides an example of the latter. Beginning operation in 1966 with a peak power output much greater than the facility at Strangford, 240 MW supplied by 24 turbines, the Rance system produces electricity at an annual level of 600 GWh, comparable to the output of a similarly sized coal-fired plant but without the emissions of CO_2 associated with the latter.

A summary of the major tide-exploiting power plants installed and planned globally is presented in Table 13.1 (http://en.wikipedia.org/wiki/List_of_tidal_power_stations). The total capacity of the plants identified as operational amounts to just a little more than 500 MW. Accounting for the higher value quoted for the capacity of the facility under construction at Inchon in South Korea, the capacity available globally could increase to as much as 1.8 GW by 2017. To put this in context, China adds approximately 2 GW of coal-fired power-generating capacity each week to respond to that country's increasing demand for electricity. Tidal energy represents a minor source of electric power at present. While it may grow in the future, especially in regions where the resource is particularly favorable, for example, in the United Kingdom or in South Korea, it is unlikely to

TABLE 13.1

Tidal Power Plants in Operation or under Construction as of August 2010

Plant	Capacity (MW)	Country	Data Operational
Sihwa Lake	254	South Korea	2011
Rance	240	France	1966
Annapolis Royal Bay of Fundy	20	Canada	1984
Jiangxia	3.2	China	1980
Kislaya Guba	1.7	Russia	1968
Uldolmok	1.5	South Korea	2009
Strangford Lough	1.2	Northern Ireland	2008
Incheon	818–1,320	South Korea	Under construction, proposed operational in 2017

Source: http://en.wikipedia.org/wiki/List_of_tidal_power_stations.

make a significant contribution to the world's demand for carbon-free, renewable energy at any time in the foreseeable future.

KEY POINTS

1. Temperature increases by an average of 25°C per kilometer below the Earth's surface.
2. The increase is greatest in regions that are tectonically active, notably in the western United States and on the edges of the Pacific Ocean.
3. The increase of temperature with depth results from release of energy associated with decay of radioactive elements in the crust combined with transfer of energy by convection and conduction from the underlying mantle and core.
4. In selected regions, contact with hot lava of water percolating from the surface combined with geological conditions favoring subsequent entrapment of this water leads to formation of what are known as hydrothermal reservoirs.
5. Drilling into these reservoirs provides a source of hot water or steam that can be used to generate electricity at the surface. The hot water or steam can be available as an intrinsic property of the reservoir, or it can be generated by piping water into the reservoir from the surface.
6. Geothermal energy, extracted from hydrothermal reservoirs, accounts for approximately 0.33% of current US electricity-generating capacity, the bulk (96%) located in California (81%) and Nevada (15%).

7. A study by the Idaho National Laboratory of the US Department of Energy suggests that with a modest investment in research on what is referred to as enhanced geothermal systems (EGS) this source could cost competitively account for as much 10% of US electricity demand in 2050.
8. EGS envisages piping water into hot rock fractured either naturally or artificially to permit sufficient contact between the water and the hot rock to raise the temperature of the water to the level that would allow for economically competitive production of electricity.
9. Because EGS would not be restricted to exploiting hydrothermal reservoirs, the opportunity to mine geothermal energy to produce electricity could be expanded to the point where geothermal sources could make an important contribution to future demand for electricity not just in the United States but also globally.
10. Opportunities for economically and environmentally constructive future development of tidal energy for electric power production are limited. Despite enthusiasm for prospects in specific regions, it is likely to continue to represent at best a minor global source.

References

Geothermal Energy Association (GEA). 2013a. *2013 annual US geothermal power production and development report.* Washington, D.C.: Geothermal Energy Association.

Geothermal Energy Association (GEA). 2013b. *2013 geothermal power: International market overview.* Washington, D.C: Geothermal Energy Association.

Geothermal Energy Association (GEA). 2014. *2014 annual U.S. and global geothermal power production report.* Washington, DC: Geothermal Energy Association.

Idaho National Laboratory (INL). 2006. *The future of geothermal energy impact of enhanced geothermal systems (EGS) on the United States in the 21st century.* Idaho Falls, ID: Idaho National Laboratory.

14

Plant Biomass as a Substitute for Oil in Transportation

AS DISCUSSED IN Chapter 3, the transportation sector accounts for approximately a third of total emissions of CO_2 in the United States, with a smaller fraction but a rapidly growing total in China. Combustion of oil, either as gasoline or diesel, is primarily responsible for the transportation-related emissions of both countries. Strategies to curtail overall emissions of CO_2 must include plans for a major reduction in the use of oil in the transportation sector. This could be accomplished (1) by reducing demand for transportation services; (2) by increasing the energy efficiency of the sector; or (3) by transitioning to an energy system less reliant on carbon-emitting sources of energy. Assuming continuing growth in the economies of both countries, option 1 is unlikely, certainly for China. Significant success has been achieved already in the United States under option 2, prompted by the application of increasingly more stringent corporate average fuel economy (CAFE) standards. And the technological advances achieved under this program are likely to find application in China and elsewhere, given the global nature of the automobile/truck industry. The topic for discussion in this chapter is whether switching from oil to a plant- or animal-based fuel could contribute to a significant reduction in CO_2 emissions from the transportation sector of either or both countries, indeed from the globe as a whole. The question is whether plant-based ethanol can substitute for gasoline and whether additional plant- and animal-derived products can cut back on demand for diesel. The related issue is whether this substitution can contribute at acceptable social and economic cost to a net reduction in overall CO_2 emissions when account is taken of the entire lifecycle for production of the nonfossil alternatives.

There is an extensive history to the use of ethanol as a motor fuel. Nicolas Otto, credited with the development of the internal combustion engine, used ethanol as the energy source for one of his early vehicle inventions in 1860. Henry Ford designed his first automobile, the quadricycle, to run on pure ethanol in 1896. While gasoline became the motor fuel of choice in the 1920s, there were more than 2,000 fueling stations in operation in the US Midwest in the 1930s marketing a mixture of gasoline blended with 6% to 12% ethanol. Use of ethanol declined subsequently only to revive in the 1970s in Brazil and later in the 1980s in the United States. The initiatives in both cases were taken not with the goal of reducing emissions of CO_2 but rather in response to the volatility of prices for oil in the wake of the oil crises of the 1970s and, specifically for the United States, concern as to the reliability of supplies at a time when the country was becoming increasingly dependent on imports to meet its accelerating demand. The feedstock of choice for ethanol in the United States was corn; in Brazil, sugar cane.

The chapter begins with an account of the current status and future prospects for production of ethanol from corn and cellulose, continuing with a discussion of the source from sugar cane, followed by analysis of the prospects for biodiesel. It concludes with a summary of key points.

ETHANOL FROM CORN

Some 13.3 billion gallons of ethanol were produced from corn in the United States in 2013, a decrease from the record of 13.93 billion gallons produced in 2011 but an increase of more than 800% with respect to the level realized in 2000. The drop in production in 2012 was occasioned by a serious drought in the corn-producing region of the United States. Production of corn fell from a level of 10.99 billion bushels in 2011 to 10.4 billion bushels in 2012 (a bushel consists of 25.4 kg or 56 pounds of kernels with a moisture content of about 15% composed of approximately 78,800 kernels). Production of ethanol declined accordingly. The recent history of ethanol production and consumption and the balance between imports and exports is summarized in Table 14.1.

The Renewable Fuel Standard (RFS) enacted in the United States through the Energy Policy Act of 2005, expanded and extended through the Energy Independence and Security Act of 2007, mandates the blending of renewable fuels with conventional transportation fuels in the United States in increasing amounts, building up to 36 billion gallons in 2022. To place these numbers in context, the United States currently consumes approximately 130 billion gallons of gasoline on an annual basis together with 50 billion gallons of diesel. The mandate is administered through a program of renewable identification numbers (RINs) overseen by the US Environmental Protection Agency (EPA). RINs are issued at the point of production or import of the renewable fuel and are intended to provide a measure of the savings in greenhouse gas emissions that could be realized from use of specific renewable fuels. RINs can be traded or banked in any particular year to offset shortfalls in blending of renewables in subsequent years. With the drop in domestic production of ethanol in 2012, the refining industry in the United States failed to meet the mandate assigned for that year. It was able to compensate by

TABLE 14.1

Ethanol Production, Consumption, and the Balance between Imports and Exports in the United States (billions of gallons)

Year	Production	Net Imports	Stock Change	Consumption
1990	0.75	Not available	Not available	0.75
2000	1.62	1.62	−0.03	1.65
2007	6.52	0.44	0.07	6.89
2008	9.31	0.53	0.16	9.68
2009	10.94	0.20	0.10	11.04
2010	13.30	−0.38	0.06	12.86
2011	13.93	−1.02	0.01	12.89
2012	13.30	−0.25	0.10	12.95
2013	13.31	−0.32	−0.18	13.18

Source: US EIA (http://www.eia.gov/tools/faqs/faq.cfm?id=90&t=4).

importing sugar cane–derived ethanol from Brazil and by cashing in RINs banked from previous years.

The growth of the US ethanol industry was supported by a tax subsidy dating back to the 1970s: beginning at 40 cents for each gallon of ethanol blended with gasoline, peaking at 60 cents a gallon in 1984, and dropping more recently to 45 cents. The subsidy is estimated to have cost the US taxpayer as much as $6 billion in 2011. The industry was further supported by a tariff of 54 cents per gallon imposed on ethanol imports. Congress eliminated both the subsidy and the tariff at the end of 2011 as part of broad-based actions adopted to reduce the federal deficit. The industry, surprisingly, offered little opposition to the steps taken with respect to ethanol. *The New York Times* (http://www.nytimes.com/2012/01/02/business/energy-environment/after-three-decades-federal-tax-credit-for-ethanol-expires.html) quoted Matthew A. Hartwig, a spokesperson for the Renewable Fuels Association, an industry trade association, as follows: "We may be the only industry in US history that voluntarily let a subsidy expire. The marketplace has evolved. The tax subsidy is less necessary now than it was just two years ago." Mr. Hartwig's reaction would likely have been different had Congress chosen at the same time to eliminate the mandate for renewable fuels. It is this mandate that guarantees, at least for the immediate future, the continuing prosperity of the domestic US ethanol industry.

Ethanol belongs to the class of chemical compounds known as the alcohols. Molecules in this family are characterized by an OH group bonded to a hydrocarbon backbone. Ethanol (CH_3CH_2OH), for example, is equivalent to ethane (CH_3CH_3) with one of the hydrogen atoms replaced by OH. If you imbibe beer, wine, or spirits, ethanol is the ingredient that gives your drink its alcoholic punch. It is a liquid at room temperature with a boiling point of 78°C. The energy content of a gallon of ethanol is approximately two-thirds that of a gallon of gasoline (76,000 Btu as compared to 115,000 Btu). What this means is that to replace the energy content of a gallon of gasoline, one would need approximately 1.5 gallons of ethanol (remember, a gallon is a measure of volume, not of energy).

As of September 5, 2014, wholesale US prices per gallon for gasoline and ethanol were \$2.87 and \$2.01, respectively. On an equal energy basis, ethanol was marginally more expensive than gasoline. Prices fluctuate in response to different market forces: the price of oil in the case of gasoline, corn in the case of ethanol. Trends in gasoline, ethanol, and corn prices over the past 31 years are displayed in Figure 14.1. As indicated, the difference between energy-adjusted costs for ethanol and gasoline has narrowed recently. Whether this trend will persist in the future will depend in large measure on the future course of prices for oil and corn. Since the underlying market forces governing these prices are different, there is little confidence in predicting how this trend should develop in the future.

The first step in conversion of corn to ethanol involves separation of the starch from other components of the corn. Starch, defined by the chemical formula $C_6H_{10}O_5$, accounts for a little more than 60% of the total mass of a corn kernel. Glucose ($C_6H_{12}O_6$) is produced from starch by hydrolysis—by adding a water molecule (H_2O) to each unit of the starch. Ethanol is formed subsequently by microbially mediated fermentation in which one molecule of glucose is converted to two molecules of ethanol accompanied by two molecules of CO_2. The initial fermentation process leads to an ethanol/water mixture with an ethanol content of about 8%. A sequence of as many as three distillation steps are required to produce ethanol at a concentration of 95%, with further processing needed to produce essentially pure ethanol (99.5%). Before shipping, the ethanol is

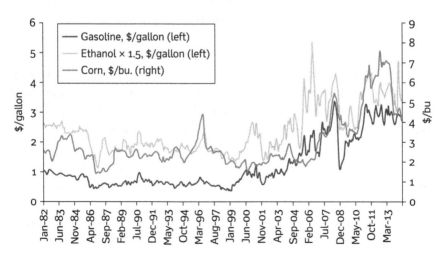

FIGURE 14.1 Historical wholesale prices for corn, ethanol, and gasoline in the United States. As indicated, the per gallon price for ethanol was increased by 50% in order to express gasoline and ethanol prices on an energy equivalent scale. (Corn data are from http://www.farmdoc.illinois.edu/manage/uspricehistory/us_price_history.html; ethanol prices and gasoline prices are from http://www.neo.ne.gov/statshtml/66.html)

denatured (made unsuitable for drinking) by addition of 5% gasoline. Otherwise, the product would be subject to the higher taxes associated with the beer/wine/spirit market. Under current manufacturing conditions in Iowa, approximately 2.8 gallons of ethanol are produced per bushel of corn (http://www.ams.usda.gov/mnreports/nw_gr212.txt), together with approximately 15 to 17 pounds of co-products that can be marketed as animal feed.

All of the steps involved in the production of ethanol from corn are energy intensive. Approximately 30% of the energy is associated with planting, growing, and harvesting the corn (we exclude in this context the sunlight absorbed to grow the corn), the balance with conversion of the corn to ethanol and subsequent distribution of the ethanol product. A number of authors (Shapouri and McAloon 2004; Pimentel and Patzek 2005; Farrell et al. 2006) have sought to determine the overall energy balance of the corn/ethanol process, to determine how the energy consumed in producing the ethanol compares with the energy delivered eventually in the product. Farrell et al. (2006) argued that the balance is positive by between about 20% and 30%; that is, the energy incorporated in the ethanol is 20% to 30% greater than the fossil energy consumed in its production. They concluded in their Ethanol Today model that the bulk of this extraneous fossil energy is supplied in the United States in the form of coal (51%) and natural gas (38%), with a relatively minor contribution from oil (6%).

Consumption of the fossil fuels involved in the corn/ethanol process is associated inevitably with emissions of CO_2. Application of nitrogen fertilizer in growing the corn results in emission of a second greenhouse gas, nitrous oxide (N_2O), which, on a molecule-per-molecule basis, is 296 times more effective as a greenhouse agent than CO_2 (Intergovernmental Panel on Climate Change 2001). Allowing for emissions of N_2O in addition to CO_2, Farrell et al. (2006) concluded that the climate impact of a corn-based ethanol–gasoline substitution under current farming and manufacturing conditions in the United States was only marginally superior to simply burning gasoline in the first place in terms of its climate impact, a 13% reduction in greenhouse gas emissions according to their Ethanol Today model.

The United States is by far the world's largest producer of corn. Approximately 20% of the crop is exported annually, contributing close to $13 billion, at current prices, to the nation's annual trade balance. Animal feed and ethanol production account for the bulk of domestic consumption, with a much smaller fraction processed for use by humans. Corn accounts for more than 90% of the grain fed to animals in the United States (cattle, hogs, and poultry). Corn-based products consumed by humans include corn oil used in cooking, a variety of sweeteners employed in the soft drinks industry, and margarine (which I grew up calling funny or phony butter). Approximately 40% of the crop is deployed to produce ethanol.

The corn crop has more than doubled over the past 30 years with much of the increase driven by the demand for ethanol, as indicated in Figure 14.2. The government-imposed mandate for ethanol has been responsible in large measure not only for the increase in

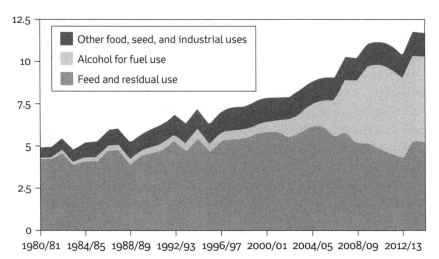

FIGURE 14.2 The US Department of Agriculture report on recent trends in US corn use. (*Source:* http://www.ers.usda.gov/topics/crops/corn/background.aspx#.VBIRcc2Pae)

production but also, to a significant extent, for the concomitant rise in prices for corn, as indicated in Figure 14.1.

Elevated prices for corn have broad implications for the affordability of a variety of food stocks not only in the United States but also around the world. As an example of the interconnection, the price of corn, as indicated in Figure 14.1, more than doubled in 2007, from $2 to $5 per bushel before settling back to about $3 per bushel in postrecession 2009, only to rise again to more than $7 in 2012. James Conca commented on the implications of the 2007 ethanol-driven rise in corn prices in an article published by *Forbes Magazine* in April 2014. Under the provocative title "Its Final—Corn Ethanol Is of No Use" (http://www.forbes.com/sites/jamesconca/2014/04/20/its-final-corn-ethanol-is-of-no-use/?&_suid=14105561380880682219246402382 9), he wrote:

> In 2007, the global price of corn doubled as a result of an explosion in ethanol production in the U.S. Because corn is the most common animal feed and has many other uses in the food industry, the price of milk, cheese, eggs, meat, corn-based sweeteners and cereals increased as well. World grain reserves dwindled to less than two months, the lowest level in over 30 years.

Given the mandate for a more than doubling of production of renewable fuels over the next 8 years as required by the 2005 and 2007 legislation, the impact of the problems highlighted by Conca can only increase in importance in the future.

The commitment of cropland to growing corn in the United States grew by more than 13 million acres between 2006 and 2011, to the present level of close to 100 million acres.

Over the same time frame, acreage devoted to cultivating wheat, oats, and sorghum decreased by 2.9 million acres, 1.7 million acres, and 1 million acres, respectively. Much of the expansion in acreage allocated to corn occurred at the expense of grassland and prairie, exploiting resources that were markedly less suitable for corn production, requiring extensive irrigation and higher applications of fertilizer. All of these developments may be attributed directly to the mandate for increased production of renewable fuels. The legislation was intended primarily to address the vulnerability the nation felt due to its increasing dependence on imported oil. Given the increase in domestic production of oil and gas in the interim resulting from development of the shale resource and the unintended consequences of the mandate in terms of food prices, it may be time to repeal the legislation that defined the mandate in the first place. A bill filed in the US Senate in December 2013 by Democratic Senator Diane Feinstein and Republican Senator Tom Coburn, supported by eight of their colleagues, could accomplish this objective. Not surprisingly, the farm lobby has aggressively opposed this measure.

ETHANOL FROM CELLULOSE

As discussed, the problem with focusing on corn for a major expansion in production of ethanol relates to the competition it triggers between deployment of the resource for fuel as opposed to its use as a food source for humans and animals. The conflict could be resolved if the ethanol were produced from plant cellulose rather than corn.

Cellulose, composed of long chains of glucose molecules, is the most abundant component of the biosphere. Together with lignin, it is responsible for the structural integrity of plants—trees and grasses. It is the lignin and cellulose that compose the trunks and branches that allow trees to stand tall. Cellulose cannot be digested directly by either humans or by most animals. Ruminants (cattle, sheep, and goats), however, are able to live on cellulose. They do so, however, through a symbiotic relationship with bacteria present in their stomachs (rumens). The plant material the rumen ingests is processed and decomposed by the bacteria, providing nutrients for the animal host while at the same time delivering nourishment to the bacteria. This is what is meant by a symbiotic relationship: both parties benefit.

Several steps are required to extract the sugars from cellulose that can be fermented to produce ethanol. First, the cellulose must be separated from the lignin. The preferred method for doing this involves first breaking down the plant material into smaller units, pellets measuring no more than a few millimeters. The pellets are exposed to either a dilute acid at high temperature (higher than 230°C) or to a more concentrated acid at lower temperature (100°C). The large surface area of the pellets relative to their volume promotes the ability of the acid to isolate the cellulose. Further processing is required to release the fermentable sugars. A key objective of current research is to find an economical means to accomplish this objective. A more extensive discussion of possible approaches is presented in McElroy (2010).

Should the cellulose option provide an important source of ethanol for the United States in the future, switchgrass, a perennial plant native to the American prairies, could provide a critical source for the required cellulose (perennial means that, after mowing, the grass will regrow; it does not need to be replanted). Since it is native, switchgrass is expected to be relatively pest resistant. Furthermore, its demand for fertilizer, specifically energy-expensive nitrogen, is significantly less than that for corn, by as much as a factor of 2 or 3 (NRDC 2004; Pimentel and Patzek 2005). It follows that the fossil energy requirement for growing switchgrass should be significantly less than that for corn. Emission of climate-impacting greenhouse gases from production of the cellulosic feedstock should be reduced accordingly, by as much as 83%, according to Farrell et al. (2006). Additional fossil fuel will be required, though, to transport the cellulosic feedstock to processing plants (relative to the costs for transporting corn), given the lower density (lower mass per unit volume) of the cellulosic material. In the absence of detailed information on the processing protocol, it is difficult to predict either the composite energy efficiency of the ethanol option or the overall impact on emissions of greenhouse gases. Further options to provide the feedstock needed to supply the cellulosic option could include fast-growing trees such as poplars and willows, surplus and waste products such as discarded paper, sawdust, and the stocks of crops for which there is currently minimal economic demand (Lave et al. 2001; NRDC 2004).

Three factories/distilleries are currently under advanced stages of development for production of ethanol from cellulose in the United States—one in Kansas (in Hugoton), the other two in Iowa. The plant in Hugoton is owned and operated by the Spanish company Abengua; the plants in Iowa are owned and operated by US companies POET-DSM Advanced Fuels and DuPont. The facility in Hugoton was supported by a guaranteed loan of $132.4 million from the US Department of Energy (DOE). The POET-DSM plant in Emmetsburg, Iowa, which went into operation on September 3, 2014, benefitted from grants totaling $122.6 million from a combination of DOE ($100 million), the State of Iowa ($20 million), and the US Department of Agriculture ($2.6 million). These plants have the capacity individually to produce a little more than 25 million gallons of ethanol annually, processing approximately 350,000 tons of corn-derived biomass, an output that may be compared with the 13.3 billion gallons produced in 2013 by the more than 200 corn-based facilities that were then operational in the United States.

The Renewable Fuels Standard (RFS) defined in the Energy Independence and Security Act of 2007 (EISA) called for 500 million gallons of cellulosic ethanol to be available in 2012, rising to 16 billion gallons in 2022. A breakdown of the mandate as it currently exists is presented in Figure 14.3. In December 2011, the Environmental Protection Agency (EPA) lowered the requirement for cellulosic ethanol in 2012 to 10.45 million gallons, responding to the fact that despite the growth of the industry, it had failed to meet the ambitious goals set by EISA. The US Court of Appeals for DC vacated the EPA mandate for 2012 in January 2013. On February 28, 2013, EPA retroactively dropped the mandate for 2012 to zero. The original target for cellulosic ethanol

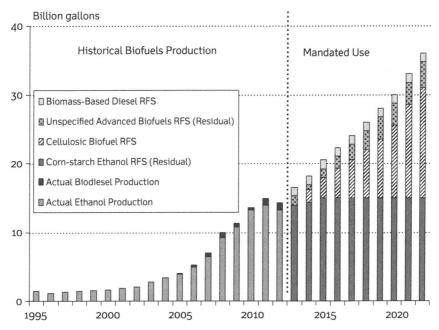

FIGURE 14.3 Renewable fuels standard (RFS2) versus US biofuel production since 1995. (*Source:* Schnepf and Yacobucci 2013)

for 2013 was set at 1 billion gallons. The EPA has since revised this requirement down to 14 million gallons. At the same time, the Agency placed a cap of 15 billion gallons on corn-based ethanol effective in 2015 and for subsequent years while maintaining the overall goal set by EISA—for production of 36 billion gallons of renewable fuels by 2022.

The prospects for cellulosic ethanol are not promising. The RFS, as currently defined, projects that production of ethanol from cellulose should climb to about 16 billion gallons by 2022, slightly higher than the target now set for production from corn. Despite generous subsidies from government, the cellulosic industry has failed to meet the mandate set by EISA in 2007. It is likely that the EPA will have to further reduce the target defined in the original legislation. The mandate for renewable fuels legislated by EISA contributes to an inevitable distortion of market forces affecting costs and supplies for all forms of renewable fuels. A case can be made to encourage at least a minimum production of ethanol given its importance as an antiknock additive to gasoline—an environmentally preferable alternative to methyl tertiary–butyl ether (MBTE), which has been phased out more recently in the United States in response to concerns as to the product's persistent carcinogenic contamination of ground waters. The current level of production of ethanol from corn is more than sufficient to meet this objective. My conclusion: the future of cellulosic ethanol should be determined by the ability of the product to compete on a price basis with alternatives absent subsidies (including corn and the sugar cane–derived products, the latter discussed later), while attaching at least some weight to its potential to reduce net emissions of climate-altering greenhouse gases.

ETHANOL FROM SUGAR CANE

Brazil trails only the United States in terms of production of ethanol. While corn provides the dominant feedstock for ethanol production in the United States, sugar cane is the primary source in Brazil. China, the world's third largest producer, elected to use cassava, sweet potato, and sorghum rather than grain as source for its biofuel. As indicated earlier, the United States produced 13.3 billion gallons of ethanol in 2013. Brazil ranked number 2 with 7.4 billion gallons, followed by China (6.1 billion gallons), the European Union (1.3 billion gallons), India (545 million gallons), and Canada (523 million gallons). The rest of the world accounted for an additional 727 million gallons.

Production of ethanol stalled in Brazil between 2011 and 2012, reflecting a combination of poor harvest conditions and government policies designed to keep gasoline prices artificially low compared with global market conditions. It has recovered since in response to policies requiring a minimum blend of 25% (E25) for all gasoline sold in the country.

The program to promote production and use of ethanol in Brazil began in the 1970s under the auspices of the then military government of the country. The motivation was the same as what prompted the initiatives adopted by the United States in the 2000s—to protect against uncertainties in prices and supplies of oil on the international market. In 1970, the Brazilian government ordered that all gasoline sold in Brazil should be blended with 10% ethanol (E10), a percentage that was increased subsequently to 20% and as indicated more recently to 25%. Cars capable of running on either ethanol or gasoline or a combination of both—so-called flex fuel vehicles—were introduced in the late 1970s. Today, more than 80% of all nondiesel cars operational in Brazil are flex fuel.

On an energy basis, sucrose accounts for approximately 30% of the photosynthetic product of sugar cane, the balance contributed more or less equally by the leafy material of the plant and by bagasse, the fibrous component that contains the sucrose. After harvesting, cane in Brazil is transported to one of several hundred ethanol distilleries scattered across the cane-producing regions of the country, concentrated in the south near Sao Paulo and in the northeast. As a first step, the cane is crushed through a series of rollers much like the rollers that were used in the past to ring water from clothes after washing. The juice extracted in this manner consists of a mixture of water and sucrose. The sucrose is concentrated by boiling off the water and is converted subsequently to ethanol by fermentation following procedures similar to those used to process the sugar produced from corn. Burning the bagasse left behind when the sucrose is separated from the cane supplies the energy needed to boil off the water in the cane juice and to fuel subsequent fermentation and concentration of the ethanol. The electricity requirements of the distilleries are met typically by burning the bagasse that remains, producing the steam required to operate conventional power-generating turbines, often with a surplus of electric power that can be fed to the grid. Fossil energy is consumed in producing the nitrogen fertilizer used to grow the sugar cane, in harvesting it, and in transporting it to the distilleries. Emission of greenhouse gases from the combination of all these activities is more than offset, however, by the savings associated with the eventual substitution of the sugar cane–based ethanol for gasoline in the transportation system. All in all, the procedures implemented to produce ethanol from sugar cane in Brazil are positive not only in terms of their energy efficiency but also in the context of the contribution they make to reducing emissions of climate-altering greenhouse gases.

BIODIESEL

Plant-based oils provide the dominant input for current production of biodiesel—soybeans, rapeseed, canola, palm, cottonseed, sunflower, and peanut—with additional contributions from animal fats and recycled grease (used cooking oil). The key ingredients of both the plant-based oils and animal fats are chemicals belonging to the family of the triglycerides. The first step in producing biodiesel involves reacting the bio-sourced oil or fat with an alcohol such as methanol or ethanol, resulting in a mixture of esters and glycerol. The glycerol is then removed. The ester that remains constitutes the basic component of the resulting biodiesel.

Biodiesel consists of a long chain of carbon atoms with attached hydrogen atoms. It differs from oil-based diesel largely in the presence at one end of the hydrocarbon chain of an ester functional group formed as a result of the reaction with the acid. The ester group consists of a carbon atom linked to the hydrocarbon chain, double bonded to an oxygen atom, single bonded to a second oxygen atom, as illustrated in Figure 14.4. Biodiesel differs from oil-based diesel, also depicted in Figure 14.4, primarily in the absence of the ester group in the latter case.

As indicated in Table 14.2, the United States, with 1.3 billion gallons, led the world in terms of production of biodiesel in 2013. Germany was in second place with 820 million gallons, followed by Brazil (770 million gallons) and Argentina (610 million gallons).

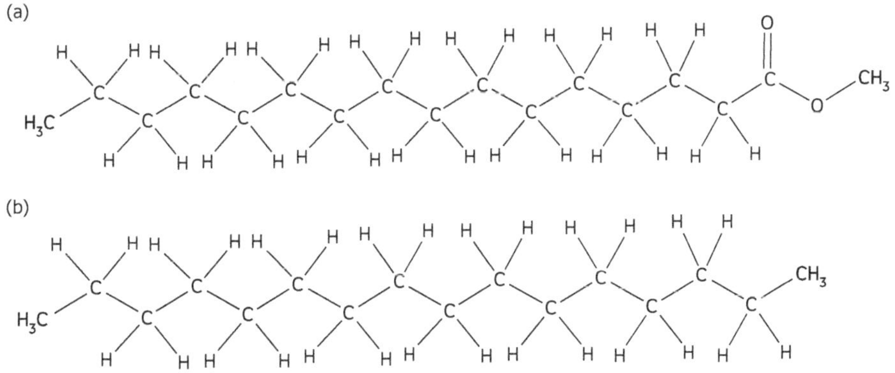

FIGURE 14.4 Structure of typical (a) biodiesel- and (b) oil-based diesel molecules. (*Source:* adapted from http://www.goshen.edu/chemistry/biodiesel/chemistry-of/)

TABLE 14.2

Production of Biodiesel for 2013 by Country
Production (million gallons)

Rank	Country	Production (million gallons)
1	United States	1,268
2	Germany	819
3	Brazil	766
4	Argentina	608
5	Indonesia	528
6	France	528
7	Thailand	291
8	Singapore	246
9	Poland	238
10	Colombia	159
11	The Netherlands	106
12	Australia	106
13	Belgium	106
14	Spain	79
15	Canada	53
16	China	53

The United States accounted for 21% of total global production in 2013. Europe, where diesel-driven cars are more popular than in the United States, was responsible for 32% of the global total.

Production of biodiesel in the United States increased in 2013 by 35% relative to 2012, driven in part by a $1.00 per gallon blending tax credit that expired on December 31, 2013. A further consideration was the adjustment in the RFS requirement for biodiesel—an increase from 1 billion gallons in 2012 to 1.28 billion gallons in 2013. The United States switched from being a net exporter of biodiesel in 2012 to a net importer in 2013. An important source of these imports came from Argentina, which was obliged to reduce its exports to Europe in 2013 as a result of an antidumping duty imposed by the European Union. Given the elimination of the tax credit, constraints imposed by limited supplies and potentially elevated prices for soybean feedstock, and the limited capacity for increased domestic production of biodiesel, the US Energy Information Administration (EIA) concluded that the United States may be forced to rely increasingly on imports to meet future mandated requirements for this product (http://www.eia.gov/todayinenergy/detail.cfm?id=16111).

Biodiesel is superior to oil-based diesel in terms of emissions of hydrocarbons, carbon monoxide (CO), sulfates, and particulates, advantages that are offset by a minor increase in emissions of nitrogen oxides. Furthermore, the energy content of biofuels per gallon is superior to that of ethanol. There are valid reasons to promote at least a limited production of biodiesel, especially from sources that could contribute to a net reduction in emission of greenhouse gases. Oil-rich algae cultivated in conjunction with wastewater treatment facilities merit attention as a potential environmentally friendly future source. On the other hand, extraction of oil from palm plants could have an insidious impact should cultivation in this case require displacement of high-carbon-content tropical forests, which harbor much of the Earth's biodiversity. What is needed is a detailed lifecycle analysis of the overall production process, from cultivation and harvesting, to conversion of feedstock to fuel, to eventual delivery of product to market.

It may be time to drop the mandate for specific biofuels imposed by the RFS in the United States and by similar legislation in Europe. The overriding goal should be to reduce the impact of emissions from the transportation sector on the climate system. There may be better and more effective means to accomplish this objective.

KEY POINTS

1. The Energy Policy Act of 2005, extended and expanded through the Energy Independence and Security Act of 2007, set ambitious standards for future production and consumption of renewable fuels in the United States. The legislation calls for 36 billion gallons to be available by 2022, equivalent to 20% of current consumption of gasoline plus diesel.
2. The EPA is charged with setting annual standards for individual fuels, adjusting requirements where necessary to account for existing market conditions, with an overarching goal to reduce emissions of climate-altering greenhouse gases.
3. Ethanol produced from corn is the dominant current source of renewable fuel in the United States, accounting for 13.3 billion gallons in 2013. Responding to concerns that demand for corn as a feedstock for ethanol is negatively impacting prices for the resource as food for humans and animals while distorting supplies and prices for other agricultural commodities, the EPA has imposed a cap of 15 billion gallons on future annual production of ethanol from this source.
4. The Renewable Fuel Standard (RFS) in the United States called for production of 500 million gallons of ethanol from inedible plant-derived cellulose by 2012, rising to 16 billion gallons by 2022. The requirement was lowered to 14 million gallons for 2013. It is unlikely that the long-term goal for the cellulosic source can be realized and that further downward revisions in the mandate will be necessary.
5. Brazil accounted for 7.4 billion gallons of ethanol from sugar cane in 2013, second only to the source from corn in the United States, with, on a per

gallon basis, significantly lower emissions of greenhouse gases. Production of sugar competes with ethanol in demand for sugar cane in Brazil. Since Brazil is a major supplier of sugar to global markets, the emphasis on ethanol production has implications for both global supplies and prices for sugar, a further example of the conflict between growing crops for fuel versus food. China was the third largest producer of ethanol in 2013, using cassava, sweet potato, and sorghum as feedstock rather than grain.

6. Plant-based oils and animal fats accounted for worldwide production of 6.2 billion gallons of biodiesel in 2013, with the United States responsible for 21% of the total. Demand for diesel is particularly high in Europe, which accounted in the same year for 32% of global consumption of biodiesel. A concern is that production of biodiesel from palm oil could contribute to an expansion of plantations into tropical forested regions with consequent release of large quantities of carbon stored in these environments—a potential net negative for climate.

7. It might be time to drop the mandate for specific biofuels imposed by the RFS in the United States and by similar legislation in Europe. There may be better and more effective approaches to reducing emissions of greenhouse gases from the transportation sector.

References

Farrell, Alexander E., R. J. Plevin, B. T. Turner, A. D. Jones, M. O'Hare, and D. M. Kammen. 2006. Ethanol can contribute to energy and environmental goals. *Science* 311: 506–508.

IPCC. 2001. The Intergovernmental Panel on Climate Change. *Climate change 2001: The scientific basis*. Contribution of Working Group 1 to the Third Assessment. Cambridge: Cambridge University Press.

Lave, L. B., W. M. Griffin, and H. Maclean. 2001. The ethanol answer to carbon emissions. *Science and Technology*. http://bob.nap.edu?issues/18.2/lave.html

McElroy, M. B. 2010. *Energy perspectives, problems, and prospects*. New York: Oxford University Press.

National Resources Defense Council (NRDC). 2004. *Growing energy: How biofuels can help end America's oil dependence*. http://www.nrdc.org/air/energy/biofuels.pdf

Pimentel, D., and T. W. Patzek. 2005. Ethanol production using corn, switchgrass, and wood: Biodiesel production using soybean and sunflower. *Natural Resources Research* 14, no.1, 65–76.

Schnepf, R., and B. D. Yacobucci. 2013. *Renewable fuel standard (RFS): Overview and issues*. Washington, D.C.: Congressional Research Service.

Shapouri, H., and A. McAloon. 2004. *The 2001 net energy balance of corn ethanol*. Washington, D.C.: U.S. Department of Agriculture. Also available at www.usda.gov/oce/oepnu.

15

Limiting US and Chinese Emissions

THE BEIJING AGREEMENT

IN A LANDMARK agreement reached in Beijing on November 11, 2014, US President Obama and his Chinese counterpart President Xi Jinping announced plans to limit future emissions of greenhouse gases from their two countries. President Obama's commitment was for the United States to emit 26% to 28% less carbon in 2025 than it did in 2005, a target more ambitious than one he had announced earlier (in 2009) that would have called for a decrease of 17% by 2020. The prior target would have required a reduction in emissions at an annual average rate of 1.2% between 2005 and 2020. The more recent agreement dictates a faster pace, at least in the later years, 2.3%–2.8% per year between 2020 and 2025. The longer term goal for US climate policy, announced at the Climate Change Summit in Copenhagen in 2009, is to reduce emissions by 83% by 2050 relative to 2005.

President Xi's commitment in Beijing was that China's emissions would peak by 2030, if not earlier, and that nonfossil sources would account for as much as 20% of China's total primary energy consumption by 2030. As indicated in the fact sheet released by the White House describing the agreement (http://www.whitehouse.gov/the-press-office/2014/11/11/fact-sheet-us-china-joint-announcement-climate-change-and-clean-energy-c), Xi's pledge would require "China to deploy an additional 800–1,000 GW of nuclear, wind, solar, and other zero emission generation capacity by 2030—more than all of the coal-fired plants that exist in China today and close to total current electricity generating capacity in the United States."

The key question is whether the United States and China can live up to the ambitious goals set by their presidents. The more serious challenge may be for the United States, where the President Obama's ability to influence policy is conditioned to a large extent by rulings by the Supreme Court declaring that climate-impacting emissions can be regulated under authority granted by the Clean Air Act. Obama, however, is scheduled to leave office in January 2017. The Congress that took over in January of 2015, with Republican majorities in both the House of Representatives and the Senate, is unlikely to share his view as to the importance of the climate issue. They may seek to overturn the executive actions he has taken and might take in the future to limit emissions. They are unlikely, however, to meet with success, since the President can veto any contrary legislation that might come to his desk and Congress is unlikely have the votes to override his veto. The question is what happens when Obama leaves office, when a new administration and a new Congress come into office with potentially different views as to the importance of the climate issue. Can the Obama commitment survive under these circumstances? President Xi is in the early days (late 2014) of an initial 5-year presidential term in China, renewable potentially for a second 5 years. He heads a government with significantly greater executive authority than its more fragmented counterpart in Washington. The odds suggest that China may be more likely than the United States to live up to the landmark climate agreement struck in Beijing in November of 2014.

The chapter focuses on what the Beijing agreement implies in the near and intermediate terms for energy policies in both the United States and China, concluding with a summary of key points.

THE BEIJING AGREEMENT: THE CHALLENGE FOR THE UNITED STATES

As indicated earlier (Chapter 3), the transportation and power sectors accounted for 72% of total US CO_2 emissions in 2013, 34% and 38%, respectively. Historical data on emissions from these sectors are displayed in Figure 15.1, which includes also the target that would have to be met in 2025, assuming that the decrease in emissions from the combination of these sectors should mirror the decrease envisaged for greenhouse gases as a whole. The breakdown in terms of emissions from the individual fossil sources—coal, oil, and natural gas—is presented in Figure 15.2. Emissions of CO_2 from oil and coal both peaked in 2005, at annual levels of 2,623 million tons and 2,182 million tons, respectively, dropping to 2,240 million tons and 1,722 million tons, respectively, by 2013. Emissions from natural gas increased from 1,183 million tons in 2005 to 1,399 million tons in 2013. Reductions in the use of oil, employed largely in the transportation sector, and coal, deployed primarily in the power sector, accounted for the bulk of the 10.5% reduction in emissions between 2005 and 2013.

A combination of factors was responsible for the decrease in oil use, notably a drop in the number of miles driven and a fleet of more efficient cars and trucks. The decrease in the use of coal resulted primarily from replacement of coal by natural gas as the fuel of

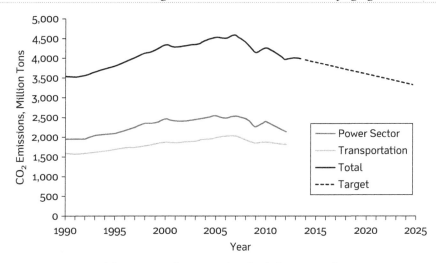

FIGURE 15.1 Historical data on annual emissions from both the power and transportation sectors with the dashed line indicating the trend required to meet the target set for emissions in 2025, assuming that the decrease in emissions from these sectors should mirror the decrease envisaged for the overall fossil carbon economy.
(*Source:* US EIA)

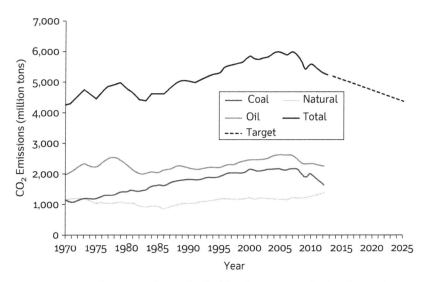

FIGURE 15.2 US annual emissions from individual fossil sources—coal, oil, and natural gas—over the period 1970 to 2012. The dashed line indicates the trend required to meet the target set for 2025.

choice in the power sector responding to unusually low prices for natural gas occasioned in recent years by the rapid increase in production from shale (demand for electricity was relatively constant over this period).

The trend in miles driven by light-duty vehicles (LDVs) is displayed for the past 33 years in Figure 15.3. The figure includes also a record of the changes in oil prices that prevailed

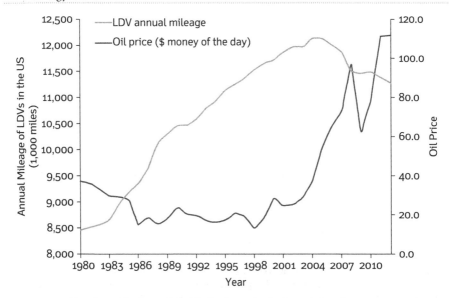

FIGURE 15.3 Trend in miles driven by US light-duty vehicles (LDVs) over the past 33 years (1980 to 2012) compared with the record of changes in oil prices ($US per barrel).
(*Source:* Davis et al. 2014; BP Statistics 2014)

over this period. There is a clear association between the number of miles driven and the price of oil. National average prices for gasoline in the United States rose first above the psychologically significant level of $2 a gallon in 2004, roughly coincident with the peak in miles driven. Gasoline prices rose above $3 a gallon in early 2008 shortly before the onset of the economic recession, settling back temporarily in the immediate aftermath before climbing again above $3, reaching a level of $3.50 in 2011.

Mileage driven is, of course, not the only factor determining emissions from the transportation sector. Important also is energy efficiency, how far the average vehicle can be driven on a given amount of fuel. Corporate average fuel economy (CAFE) standards, legislated by the US Congress in 1975 in response to the first oil shock, are designed to encourage automobile and truck manufacturers to increase the fuel efficiency of the vehicles they bring to market. The trend in CAFE standards over the past several decades and prospectively forward to 2025 is illustrated in Figure 15.4.

The initial legislation resulted in a significant improvement in the average fuel efficiency of passenger vehicles, rising from 18 miles per gallon (mpg) in 1978, to 27.5 mpg in 1985. Requirements remained relatively flat until December 19, 2007, when President Bush signed into law the Energy Independence and Security Act (EISA, discussed earlier in Chapter 14 in the context of the targets the legislation set for renewable fuels), which defined a goal of 35 mpg for the national fleet average efficiency in 2020. EISA identified different requirements for different vehicles, depending on size—stringent for small vehicles and more relaxed, yet challenging, for larger vehicles. President Obama, on July 29, 2011, announced an agreement with 13 of the largest auto manufacturing

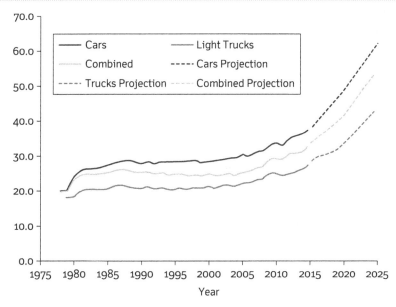

FIGURE 15.4 Trend in CAFE standards (miles per gallon gasoline equivalent) over the past several decades and prospectively forward to 2025. (http://www.epa.gov/fueleconomy/fetrends/1975-2013/420r13011.pdf)

companies to increase the average fuel economy to 54.5 mpg by 2025. The assumption in this agreement was that small cars such as the Honda Fit would realize an efficiency of 60 mpg, offsetting higher consumption, 46 mpg, by larger cars such as the Mercedes Benz S-Class. Small trucks such as the Chevy S10 would be rated at 50 mpg with larger trucks such as the Ford F-150 required to achieve 30 mpg (http://en.wikipedia.org/wiki/Corporate_Average_Fuel_Economy). The President further instructed the Environmental Protection Agency (EPA) and the Department of Transportation (DOT) to propose, by March 2015, new rules that would apply for the first time to medium and heavy trucks, requirements that would come into effect for vehicles entering the market in model year 2018.

The EPA has taken steps to limit emissions from the power sector, announcing, on September 20, 2013, standards that would apply to emissions of CO_2 from future generating facilities (http://www2.epa.gov/sites/production/files/2013-09/documents/20130920factsheet.pdf). As proposed, these regulations would distinguish between emissions from coal- and natural gas–fired systems. Fossil fuel–fired utility boilers and integrated gasification combined cycle (IGCC) plants, primarily coal-fired, would be restricted to emissions of 1,100 pounds CO_2/MWh gross over a 12-month operating period or 1,000 pounds CO_2/MWh gross over an 84-month operating period. Natural gas–fired systems would be limited to emissions of 1,000 pounds CO_2/MWh for larger units (greater than 850 million BTU/hr) and 1,100 pounds CO_2/MWh for smaller units (less than 850 million BTU/hr). The regulations would place a particular burden on coal-fired systems. Emissions from coal-fired plants currently average about 2,100 pounds CO_2/MWh as compared to 1,220 pounds CO_2/MWh for gas-fired systems (http://www.eia.gov/tools/faqs/faq.cfm?id=74&t=11).

For new coal-fired plants to meet the proposed standards, they would have to be equipped to capture CO_2 from their exhaust. This would result inevitably in a significant increase in the price for electricity generated by these systems, an increase that could be justified only if the CO_2 captured in this process could be marketed to compensate for the extra expense. While economically profitable applications for this CO_2 could be contemplated potentially for the future (possible initiatives that could accomplish this objective are discussed in the following chapter), it is clear, at least in the short term, that the requirement to capture and dispose of exhaust gas CO_2 will place new coal-fired facilities at an important disadvantage. Under current conditions, with low prices for natural gas and anticipated high expenses for the treatment of emissions from coal-fired plants, the future for new coal-fired systems would appear to be limited.

The EPA is also exploring options to limit emissions from existing power plants. The approach in this case involves negotiations with individual states to identify cost-effective means to accomplish this objective. Options under consideration include (1) reducing the carbon intensity of individual power plants by improving the efficiency with which they convert heat to electricity; (2) reducing emissions from the most polluting plants, compensating by promoting increased utilization of plants defined by lower carbon emissions; (3) increasing investment in low and zero carbon-emitting sources; and (4) employing demand-side incentives to reduce overall demand for electricity. Following extensive consultations with individual states and stakeholders, taking account of prevailing conditions and circumstances, the EPA has defined targets that individual states are expected to meet by 2030, including interim goals to be realized beginning in 2020. On a national basis, the measures proposed are intended to result in a decrease of approximately 30% in emissions of CO_2 from the power sector by 2030 relative to 2005. As envisaged, the major share of US electricity would continue to be provided by fossil sources, with coal and natural gas individually accounting for more than 30% of total national power production (https://www.federalregister.gov/articles/2014/06/18/2014-13726/carbon-pollution-emission-guidelines-for-existing-stationary-sources-electric-utility-generating-units). Assuming that the EPA's authority to regulate emissions is not constrained by future legislation, the target for the decrease in emissions of CO_2 from the power sector by 2030 would appear to be realistic and consistent with the decrease in overall greenhouse gas emissions of 26% to 28% identified for 2025 in the Beijing agreement. To meet the overall objective, however, will require a comparable reduction in emissions from the transportation sector.

The precipitous drop in oil prices that began in November 2014 and the related decrease in the national average price of gasoline—from $3.76 a gallon on July 4, 2014, to $2.30 a gallon on December 31, 2014—will make it difficult to realize this objective for the transportation sector. When gasoline prices are low, people tend to drive more, consuming more fuel, emitting more CO_2 (Fig. 15.3). Offsetting this, cars and trucks are expected to become more efficient in the future in response to the anticipated rise in CAFE standards (Fig. 15.4). It will take time, however, to see the full effect of these mandates: the turnover of the car/truck fleet in the United States is relatively sluggish, with new vehicles accounting for only about 10% of cars and trucks on the road in any given year. Furthermore, based at least on past experience, there will be a tendency for people to buy larger cars and trucks should fuel prices continue to be depressed. With exceptionally low prices for motor fuels, could this not provide an opportunity to increase taxes on gasoline and diesel fuels, both as a means to influence behavior and as a relatively painless opportunity to raise revenue? The tax could be imposed in a revenue-neutral mode with income used to reduce taxes in other areas of the economy or simply to pay down the deficit.

As indicated in Chapter 2, prices for motor fuels are exceptionally low in the United States, less than half what they are in Europe, and significantly lower than in Japan and in other major Asian countries. The base price for motor fuels should be comparable for all countries, reflecting the international price of oil. Differences across countries respond, therefore, primarily to variations in taxes. The federal tax on a gallon of gasoline has been fixed at 18.4 cents a gallon in the United States since 1993. Had it kept pace with inflation, it should have risen in the interim to at least 30 cents a gallon. Gasoline and diesel taxes in the United States are used primarily to fund repairs and for maintenance on roads and bridges and for investments in related transportation infrastructure. The Federal Highway Trust Fund, which administers these activities, is currently effectively bankrupt, facing a deficit of as much as $160 billion over the next 10 years in the absence of new sources of revenue.

Republican Senator Corker and Democrat Senator Murphy introduced a bill in the US Senate in June 2014 that would have raised the gasoline tax by 12 cents a gallon over a 2-year period, indexing future taxes to prevailing rates of inflation. Even this modest proposal failed to gain traction, a victim of the pervasive antipathy of the US political establishment to increases in taxes for any reason, however compelling the justification.

We commented earlier (again in Chapter 2) on *The New York Times* correspondent Tom Friedman's proposal that when gasoline prices were rising to nearly $4 a gallon, it was time to consider a $1 per gallon tax on gasoline to be phased in, rising by 5 cents a month, to meet this objective. His suggestion was that the revenue raised in this manner could be used to pay down the deficit. An appropriately designed gasoline tax could play an additionally important role in encouraging more conservative use of gasoline even under conditions when prices are falling. The tax could be structured to ensure a relatively constant retail price for gasoline. This could be accomplished by allowing the tax to increase as prices dropped, to decrease if they moved in the opposite direction. To be effective in the current context (to reduce or at least maintain a relatively constant level of emissions from the transportation sector in the face of falling prices), this arrangement should have been in place by early 2014. And it should have been implemented so that the tax should have risen to a level of about $1 a gallon by the end of 2014. To introduce the tax option to adjust consumer behavior now, to encourage a return of retail gasoline prices to levels that prevailed in early 2014, would be difficult. Rather than the consumer experiencing effectively constant prices at the pump, as would have been the case had the tax been introduced earlier, it would be necessary in the present instance to use the tax to drive prices higher. Given the current political climate in the United States, prospects for introduction of such a tax are bleak. The question then is whether it might be possible to compensate for a slower rate of decline, or even an increase, in emissions from the transportation sector by requiring a more aggressive reduction from the power sector. And could this reduction be implemented cost effectively with minimal impact on the overall economy?

Coal accounted for emission of 1,664 million tons of CO_2 in the United States in 2012 with an additional 1,364 million tons from natural gas and 2,255 million tons from oil. The power sector was responsible for 2,157 million tons of total emissions with an additional 1,819 million tons from transportation. To meet the target for total emissions set by the Beijing announcement will require that composite emissions in 2025 should be lower than in 2012 by 630 million tons. Coal accounted for 1,514 TWh of electricity production in 2012 with an additional 1,225 TWh from natural gas. The bulk of the contribution from gas (1,104 TWh) was from natural gas combined cycle systems (NGCCs), supplemented by a minor contribution from gas turbines deployed under peaking conditions (121 TWh). Capacity factors for coal and NGCC systems averaged 55.7% and 41.9%, respectively, in 2012. If the bulk of the reduction in emissions required by 2025 is to come from the power sector, it will be necessary to markedly reduce the contribution from coal.

As a conservative assumption, assume that demand for electricity in 2025 will be similar to what it was in 2012. A reduction in overall emissions of CO_2 by 630 million tons could be realized by effectively eliminating the contribution from coal, substituting a combination of enhanced production from natural gas, complemented by production from zero-carbon sources such as nuclear, wind, and solar. If the deficit were to be made up solely by increased use of the existing stock of NGCC plants, these plants would have to operate at an unrealistic capacity factor (CF) level of 90%. The deficit could be accommodated by adding 126 GW of new capacity to the existing NGCC stock, assuming that the integrated NGCC system could operate at a CF level of 70%, a challenging, though potentially achievable, objective. Investments in zero-carbon-emitting systems, such as wind, would reduce the demand for new gas plants. The need for new gas systems could be completely eliminated by adding 103 GW of wind capacity, assuming that the installed systems could operate at a CF of 30%. Wind capacity installed in the United States amounted to 61.1 GW as of the end of 2013. Adding a little more than 6 GW per year over the next 10 years would appear to represent a readily achievable target. Production from solar sources, investments in which hit a record level of $150 billion in 2014, increasing by 25% with respect to 2013, could make an important additional contribution to the production of zero-carbon-emitting electric power.

Exceptionally low prices for natural gas, close to $3 per MMBTU at the end of 2014, are likely to encourage substitution of gas for coal in the US power sector. This could extend the trend responsible largely for the decrease in emissions over the past decade. To meet the longer term objective of the Obama administration, to reduce emissions by 83% by 2050 relative to 2005, it would be preferable in the short term to invest in zero-carbon-emitting sources rather than expand the commitment to natural gas, despite the evident price advantages that this could entail at least for the short term. The operational life of a typical gas-fired power plant extends to at least 40 years. Contemporary investments in such systems would have implications not only for emissions immediately but for the indefinite future.

THE BEIJING AGREEMENT: THE CHALLENGE FOR CHINA

As indicated at the outset, the commitment of President Xi with respect to China's future emissions was notably less specific than President Obama's pledge for the United States. Emissions from China were projected to peak by 2030, if not sooner. The announcement did not identify a specific value for the magnitude of the emissions at that time. Zero-carbon sources were proposed to account for 20% of total primary energy consumption. Left unstated were the magnitude and composition of the fossil fuel components that would account for the remaining 80%.

China has important reasons other than concerns about climate change to cut back on the use of fossil fuels and related emissions. Sulfur and nitrogen oxides produced as byproducts of the combustion of coal and, to a lesser extent, oil and natural gas have a significant negative impact on air quality. Chemical transformations of these species in the atmosphere, responding additionally to the presence of ammonia emitted from agricultural sources, lead to production of small particles (aerosols), characteristically 2.5 micrometers (μm) or less in size, referred to collectively as particulate matter 2.5 or $PM_{2.5}$. The presence of these particles in the atmosphere contributes to the formation of haze affecting visibility, the deterioration of which has been so extreme on occasions recently that it has been difficult to discern the identity of objects separated from the viewer by as little as a few tens of meters. More serious than the implications for visibility is the impact the haze, or smog as it is identified more appropriately, can have for humans exposed to breathing its constituent chemicals. This can lead to serious respiratory problems and, for vulnerable individuals, could even be life threatening. Paradoxically, as noted earlier in Chapter 4, the presence of these particles in the atmosphere has a positive impact on the climate system in that they can offset to some extent the warming induced by the increasing concentrations of CO_2 and other greenhouse gases. This should not under any circumstances be interpreted as justification to postpone action to reduce the conditions responsible for production of this toxic mix of local and regional pollution.

Public consciousness as to the gravity of the air pollution problem in China was sparked initially by a number of major episodes and by release through social media of data taken from a $PM_{2.5}$ instrument installed on the roof of the US embassy in Beijing. The data suggested that the concentrations of $PM_{2.5}$ were significantly higher than levels reported by official Chinese government sources. The public response was immediate and critical, prompting the central government to announce an initiative to address the issue, the Air Pollution Prevention and Control Action Plan (APPCAP). In response, the number of cities in China in which $PM_{2.5}$ would be measured was more than doubled with results reported in real time on government websites. The credibility of the central government is seriously invested in the need to mitigate this high-profile problem. In the public consciousness, it clearly ranks higher than the threat of climate change.

Policy initiatives to improve the efficiency of the energy economy are not new in China. The 11th Five Year Plan, developed in 2006, expressed the goal to reduce the energy intensity of the economy—the energy required to produce a given unit of gross domestic product (GDP)—by 20% by 2010 relative to 2006. The government announced a further goal in late 2009, to reduce the carbon intensity of the economy—the quantity of CO_2 emitted per unit of GDP—by 40% to 45% below 2005 levels by 2020. The 12th Five Year Plan, released in 2011, had dual prescriptions, to reduce the energy intensity by 16% and the carbon intensity by 17% over the interval covered by the Plan. It indicated further that nonfossil sources should account for 11.4% of total primary energy by 2015.

The objective announced by President Xi in his meeting with President Obama in Beijing in November 2014, that nonfossil sources should account for 20% of total primary energy consumption by 2030, may be considered simply a logical temporal extension of the commitments identified earlier under the 11th and 12th Five Year Plans. Plans for the more immediate near term, out to 2020, were formulated and announced more recently. They call for a cap on annual coal consumption at 4.2 billion tons, natural gas to account for 10% of total primary energy supply, nuclear power capacity to rise to 58 GW, an additional 30 GW of nuclear capacity to be under construction, hydro capacity to increase to 350 GW, investments in wind systems to reach 200 GW, solar PV to increase to 100 GW, and nonfossil sources to account for 15% of total primary energy consumption, all of this by 2020. Assuming that these ambitious near-term objectives can be realized, prospects for meeting the longer term 2030 objectives announced by President Xi in the Beijing meeting with President Obama would appear to be excellent. This conclusion is consistent with results from a comprehensive analysis of potential future prospects for the Chinese energy economy by scientists at the Lawrence Berkeley National Laboratory of the US Department of Energy (Zhou et al. 2011). Their study, which extended out to 2050, explored a variety of potential paths for China's energy and carbon futures, including detailed analyses of the influence of the important demographic, economic, and social changes anticipated to develop in China over this expanded time horizon.

The Berkeley study recognizes that China's economy is currently in transition from the status classified as developing to developed. Demand for energy is particularly high during this phase. A premium is placed on investments in energy-intensive industries such as iron, steel, and cement, products from which are needed to supply the materials required for construction of the roads, bridges, buildings, railroads, ports, and airports critical for success during this phase of development. Accompanying this rapid industrial development is an equally significant change in the structure of Chinese society, reflected specifically in a mass relocation of people from the countryside to the cities. This results in additional demands for energy, for heating and cooling of new urban residences, for refrigerators and new labor-saving devices, and for transportation. Compounding the importance of these changes is the fact that the Chinese population is aging, with the total population likely to peak at some point over the next few decades. Historical data and projections for the future changes in populations of rural and urban residents are displayed in Figure 15.5.

Zhou et al. (2011) considered two models for China's energy future, a baseline continued improvement scenario (CIS) and an alternative accelerated improvement scenario (AIS). Assumptions with respect to the growth of GDP are similar for both models: 7.7% per year from 2010 to 2020, 5.9% per year from 2020 to 2030, and 3.4% per year from 2030 to 2050. The targets announced by the Chinese government for 2020 for the combined capacities of hydro, wind, and solar, 650 GW, are in fact significantly higher than the capacities envisaged for these sources in either of the Berkeley models for 2050: 535 GW for CIS, 608 GW for AIS.

Projections of the Berkeley models for total primary energy by fuel and by sector are summarized in Figure 15.6 and 15.7. The fraction of primary energy supplied by coal is

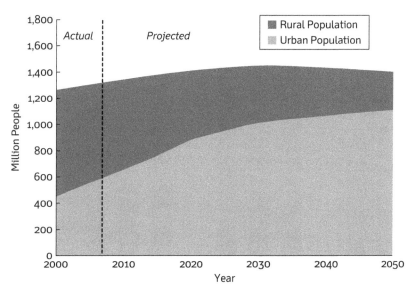

FIGURE 15.5 Historical data and projections for the future for populations of rural and urban residents.
(*Source:* Zhou et al. 2011)

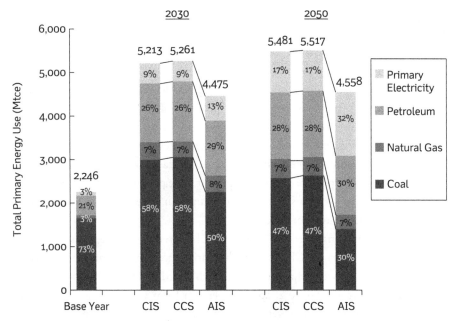

FIGURE 15.6 Projection of total primary energy use by fuel for China. CCS refers to an option that exploits carbon capture and sequestration as a strategy to reduce emissions of carbon dioxide but at a cost of more coal consumed to deliver a supply of useful energy.
(*Source:* Zhou et al. 2011)

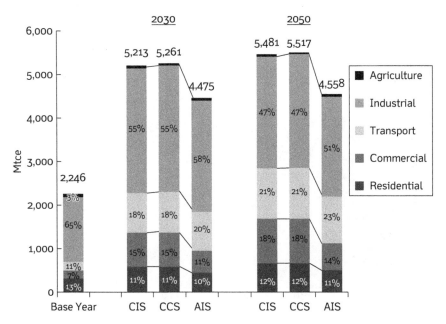

FIGURE 15.7 Projection of total primary energy use by sector in China.
(*Source:* Zhou et al. 2011)

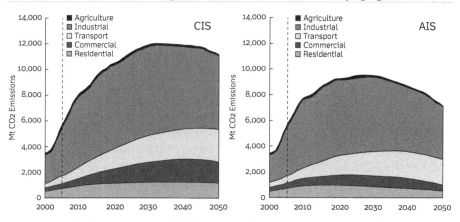

FIGURE 15.8 Carbon emissions outlook for continued improvement scenarios and accelerated improvement scenarios in China.
(*Source:* Zhou et al. 2011)

reduced from 73% in the base year (2005) to 50% in the AIS scenario for 2030, to 30% in 2050. Notably, a significant fraction of total energy continues in all scenarios to be allocated to industry, more than 50%, with the combination of commitments to the residential, commercial, and transportation sectors in the AIS model accounting for 41% of total energy use in 2030, rising to 48% in 2050.

Projected emissions of CO_2 are displayed in Figure 15.8. Emissions for both the CIS and AIS models peak around 2030. In this context, the projections are consistent with President Xi's Beijing commitment. Notably, though, even in the context of the more aggressive AIS model, emissions from China are forecast to be more than 50% higher in 2030 than they were in 2005, approximately twice the level projected for the United States in 2025. Offsetting the disparity in absolute emissions from the two countries is the fact that per capita emissions from China are currently lower than those from the United States by about a factor of 2.7 and that this difference is likely to persist.

KEY POINTS

1. The agreement between the United States and China announced by Presidents Obama and Xi in Beijing in November 2014 calls for a reduction in US emissions of CO_2 by 26% to 28% by 2025 relative to 2005.
2. Based on extrapolation of the trend in emissions observed between 2005 and 2013, this target would appear to be reasonable and readily attainable. The decrease in emissions from the power sector over this period resulted in large measure from an increase in production of power from natural gas at the expense of coal, prompted by low prices for the former. The decrease in emissions from the transportation sector was prompted by a combination of factors: more efficient vehicles responding to more demanding corporate average

fuel economy (CAFE) standards and reductions in mileage driven responding to high prices for motor fuels. In extrapolating past trends to the future it will be important to account for the impact of the precipitous drop in the price of oil that developed over the second half of 2014 and that has continued up to the present (2016).

3. The average retail price of gasoline in the United States decreased from $3.76 a gallon on July 4, 2014, to $2.30 a gallon on December 31, 2014, falling below $2.00 a gallon in early 2016. Should the lower prices persist, driving habits in the United States are likely to change, with motorists favoring larger, less fuel-efficient vehicles and using these vehicles to drive more. Emissions of CO_2 from the transportation sector are likely to rise accordingly.

4. The response of motorists to lower prices for fuel could be tempered by a tax on motor fuels. To be effective, however, this tax would need to have the potential to rise to a level of as much as a dollar a gallon. Income from the tax could be used to offset other sources of government revenue, and the tax could be administered in a revenue-neutral mode. Prospects for introduction of such a tax under current political conditions in the United States are not promising. Despite continuing improvements in CAFE standards, the decrease in emissions from the transportation sector recorded in the United States over the past decade is unlikely to persist.

5. Meeting the overall target for emission reductions announced for the United States by President Obama in Beijing will require in this case a larger contribution from the power sector. This could be accommodated by further substitution of coal-fired generation by systems fueled by natural gas and/or zero-carbon-emitting alternatives such as wind and solar. To meet the required reduction in overall emissions exclusively from the power sector, it will be necessary to effectively eliminate the contribution from coal. The preferred option for the long term would be to replace the coal source with zero-carbon-emitting alternatives rather than new natural gas–fired facilities. Investments in new gas-fired systems would have implications for emissions over the indefinite future, making it difficult to cost effectively meet the longer term objective announced by President Obama, to reduce US emissions by 83% by 2050 relative to 2005.

6. President Xi's commitment in Beijing was that Chinese emissions should peak by 2030 if not sooner and that zero carbon sources should account for as much as 20% of China's primary energy consumption by that date. The expectation is that both of these objectives can be met, noting in particular the major investments contemplated by China immediately, by 2020, in nuclear, hydro, wind, and solar facilities.

7. China surpassed the United States as the world's largest emitter of CO_2 in 2007. With continuing growth since then, it is probable that when emissions

peak as promised in 2030, the level of emissions from China will exceed those from the United States by at least 50%, 9 billion tons CO_2 a year from China as compared to a little more than 4 billion tons CO_2 a year from the United States. Per capita emissions from China, however, will be less than those from the United States by as much as a factor of 2.7.

References

BP. 2014. BP statistical review of world energy: 48. http://www.bp.com/content/dam/bp-country/de_de/PDFs/brochures/BP-statistical-review-of-world-energy-2014-full-report.pdf.

CAAC. 2013. *Air pollution prevention and control action plan* (English translation). Edited by State Council of China. Beijing, China: Clean Air Alliance of China.

Davis, S. C., S. W. Diegel, and R. G. Boundy. 2014. *Transportation energy data book*. Oak Ridge, TN: Oak Ridge National Laboratory.

US EIA. 2013. *Annual energy outlook 2013*. Washington, D.C.: U.S. Energy Information Administration.

Zhou, N., D. Fridley, M. McNeil, N. Zheng, J. Ke, and M. Levine. 2011. *China's energy and carbon emissions outlook to 2050*. Berkeley, CA: Lawrence Berkeley National Laboratory.

16

Vision for a Low-Carbon-Energy Future

THIS CHAPTER DISCUSSES steps that could be taken to realize the long-term goal of reducing, if not eliminating, climate-altering emissions associated with the consumption of coal, oil, and natural gas. I choose to focus on initiatives that could be adopted over the next several decades to advance this objective in the United States. The key elements of the vision proposed for the United States should be applicable, however, also to China and to other large emitting countries. As indicated at the outset, the overall focus in this volume has been on the United States and China, the world's largest emitters of greenhouse gases, recognizing at the same time differences in states of development and national priorities of the two countries. The vision I outline here for a low-carbon-energy future for the United States should apply also to other countries. The time scale for implementation may differ, however, from country to country, depending on details of local conditions and priorities—economic, social, and environmental.

The data presented in Chapter 3 (Figs. 3.1 and 3.2) provide a useful starting point—essential background—for discussion of potential future scenarios (US EIA 2015). They define how energy is used in the current US economy and the services responsible for the related emissions, with key data summarized in Table 16.1. Generation of electricity was responsible for emission of 2,050 million tons of CO_2 in 2013, 1,580 million tons from combustion of coal, and 442 million tons from natural gas, with a minor contribution, 34.7 million tons, from oil. The residential, commercial, and industrial sectors accounted, respectively, for 38%, 36%, and 26% of emissions associated with economy-wide consumption of electricity. The power sector was responsible for 38% of total national emissions. Transportation contributed an additional 1,826 million tons, 34% of the national total. The bulk of the emissions from transportation (98%) was associated

TABLE 16.1

Breakdowns of Energy Consumption and Associated CO_2 Emissions for the United States in 2013: (1) Residential Sector, (2) Commercial Sector, (3) Industrial Sector, and (4) Transportation Sector

	Energy Consumption (Quads)	CO_2 Emissions (million tons)
1. Residential Sector		
Electricity	4.75	785.3
Coal	0	0
Gas	5.05	268
Oil	0.893	59.9
Biomass	0.42	0
2. Commercial Sector		
Electricity	4.57	755.5
Coal	0.0454	4.28
Gas	3.36	178
Oil	0.477	33.4
Biomass	0.112	0
3. Industrial Sector		
Electricity	3.26	538.9
Coal	1.50	140
Gas	9.08	460
Oil	8.58	369
Biomass	2.25	0
4. Transportation Sector		
Electricity	0.0257	4.2
Coal	0	0
Gas	0.795	42.2
Oil	24.9	1780
Biomass	1.24	0

with consumption of petroleum products, gasoline, diesel fuel, and jet fuel, with the balance from natural gas. Natural gas deployed in residential and commercial settings—as a source among others of energy for space heating and hot water—was responsible for emissions of 268 million tons and 178 million tons, respectively, with a further 93 million tons from consumption of oil used to provide heat and hot water for buildings in winter: 60 million tons from residences, 33 million tons from commercial establishments. Emissions associated with industrial combustion of natural gas, oil, and coal accounted for 460 million tons, 369 million tons, and 140 million tons of emissions, respectively. In summary, generation of electricity was responsible for 38% of US emissions in 2013, transportation for 34%, with direct consumption of fossil fuels (oil, natural gas, and coal) in the residential, commercial, and industrial sectors contributing additionally 6%, 4%, and 18%, respectively.

As indicated in the previous chapter, the long-term commitment of the United States is to reduce net emissions of greenhouse gases by 83% by 2050 relative to 2005. The challenge is to meet this objective while maintaining the energy services and quality of life enjoyed by the current US population of 325 million, while planning for a population projected to exceed 400 million by 2050. If we are to limit the impact of increasing concentrations of greenhouse gases on future climate, we will need to markedly reduce emissions associated with the use of fossil fuels in the generation of electricity and in fueling our cars, trucks, trains, ships, and planes. And we will need to explore options for the substitution of nonfossil alternatives for natural gas and oil in the residential and commercial sectors.

We paint a picture here for a future in which energy services are delivered to an increasing extent in the form of electricity. Electricity can substitute for natural gas and oil in heating buildings, in cooking, in providing hot water, and in a variety of other applications. It can complement gasoline and diesel fuels in propelling our cars and light trucks. And it can substitute for fossil fuels in supplying the energy needed to produce the steam deployed in a variety of industrial applications. Conservation can play an important role in minimizing the overall future demand for electricity. Despite this, if we are to meet the projected expanded market, it is clear that production of electricity will have to increase, perhaps by as much as 50%. The key, though, is that the electricity must be generated with minimal emission of CO_2.

Should we persist in requiring a continuing important contribution of power from coal and natural gas, we will need to invest in equipment to capture the associated emissions of CO_2 prior to their release to the atmosphere and to either bury them in a secure depository or find a productive use for them. As noted earlier, there is an inevitable energy penalty associated with capturing, concentrating, and purifying the CO_2 included in the exhaust gases of fossil fuel–fired power plants. What that means is that it would be necessary to burn more coal and natural gas to produce a given quantity of electricity. The better option, we shall argue, is to transition to a power system less reliant on fossil sources for its energy input.

Nuclear power can provide an important source of baseload (constant, used all the time) electricity. Prospects for investment in new nuclear power facilities in the United States, at least over the immediate future, are not promising, as discussed in Chapter 9. The challenge will be to maintain as much of the present operational nuclear capacity as possible by extending the life of existing plants. The contribution of power from geothermal sources could be enhanced and could develop in the future as a significant contribution to both baseload and peaking demand. Whether this potential can be realized will depend, as discussed in Chapter 13, on investment in a significant program of research and development and a successful outcome from that activity—proof that geothermal power can provide an important source of future electricity at reasonable cost. Prospects for a consequential expansion in the contribution of power from hydro in the United States are limited (Chapter 12). The burden for a major source of future fossil carbon-free power is likely to fall thus by default on a significantly enhanced supply from wind and solar. The problem is that these sources are intrinsically variable—the wind does not blow all the time nor does the sun always shine. Much of the discussion that follows is devoted to strategies to allow maximum advantage to be extracted from these sources.

We discuss sequentially the need to build out the transmission system and opportunities to make greater use of electricity in domestic and commercial settings, combining this with a heightened emphasis on conservation, increased use of electricity in the transportation sector, and opportunities to use biomass to produce carbon-neutral, possibly even carbon-negative, products that could substitute for the fossil fuel–based sources that currently dominate the transportation energy sector.[1] We continue, addressing actions that could be considered should we fail to take the steps recommended here to prompt the transformation of the energy system required to minimize the disruption of the climate system envisaged in Chapter 4. Potential responses include initiatives to intervene actively to alter the climate to minimize the potential impact of human-induced change (what is referred to as geo-engineering) and steps to remove CO_2 from the atmosphere, to effectively cancel the input from fossil fuels. The chapter concludes with a summary of key points.

PLANNING FOR A TWENTY-FIRST-CENTURY ELECTRICITY TRANSMISSION SYSTEM

As indicated in earlier chapters, the United States is richly endowed in potential sources of renewable energy—wind, solar, hydro, and geothermal. The cost for generation of electricity from wind is already competitive with the cost for generation using alternatives, with the possible exception of natural gas, which continues to benefit from unprecedentedly low prices for the underlying commodity. Costs for production of power from distributed photovoltaic (PV) sources (installed on domestic and commercial rooftops) are declining and investments in utility-scale concentrated solar power (CSP) facilities are increasing. The challenge in capitalizing on the potential for wind and solar is that the most productive sources are physically removed from locations where demand is greatest, notably metropolitan regions in the east and west. Wind conditions are most favorable in the middle of the country, and solar sources in the southwest. Priority number 1 should be to extend the existing transmission network, to better connect favorable source regions with centers of high demand.

For much of the past century, production and distribution of electricity in the United States was controlled by a large number of small vertically integrated utilities, often single plants established to serve local markets. Through time, utilities elected to link their distribution networks, at least locally, to ensure a more reliable source of power for their customers. If generation facilities owned by one company had a problem and had to shut down, other interconnected facilities could pick up the slack. Through time, what developed in the United States were three effectively isolated distribution networks, the Eastern Interconnection, the Western Interconnection, and the Electricity Reliability Council of Texas (ERCOT) with links also to Canada, as illustrated in Figure 16.1.

The Federal Energy Regulatory Commission (FERC) has regulatory authority over interstate sales of electricity and the operation of regional markets. The North American Reliability Corporation (NERC), formed in the wake of a major blackout that struck the US Northeast in 1965, operating through eight regional entities as indicated in Figure 16.1, is charged under FERC with ensuring the orderly function of the overall national system. Notably, six of the eight reliability entities are located in the Eastern Interconnection region with the Western Interconnection and ERCOT regions operating with centralized authority as integrated units. Jurisdictions for a number of these entities (WECC, MRO, and NPCC) extend into Canada. The figure indicates also locations for the limited number of AC-DC-AC connections that currently link the Eastern Interconnection, the Western Interconnection, and ERCOT.

The Energy Policy Act of 1992 mandated open access to the transmission system. Subsequently, a variety of additional organizations, referred to as independent system operators (ISOs) or regional transmission organizations (RTOs), were created to oversee grid operations and ensure equal access to transmission services for power producers in their regions. Where wholesale markets exist and where there is a significant presence of independent power producers, responsibility for transmission planning rests largely with the RTOs or ISOs. In regions where vertically integrated utilities continue to play a major role in the overall power system, notably in the West and in Texas, the organizational structure is more complex. The distribution of RTOs and ISOs across the United States is indicated in Figure 16.2. Five of these organizations are located in the Eastern Interconnection Region (the Southwest Power Pool, the Midwest ISO, the Pennsylvania-New Jersey-Maryland Interconnection, the New York ISO, and ISO New England), one in the west (the California ISO), and one in Texas (ERCOT). The Midwest ISO extends into Canada. As indicated, there are three additional Canadian-based entities (the Alberta System Operator, the Ontario Electricity System Operator, and the New Brunswick System Operator).

To take full advantage of the country's abundant wind and solar resources will require development of a more integrated national electric grid. The present system is fragmented, reflecting the piecemeal manner in which it evolved. It will be important to invest in AC-DC-AC connections to allow power to flow freely between the three autonomous grids—the Eastern Interconnection, the Western Interconnection, and ERCOT. Conversion from AC to DC and then back to AC is necessary to ensure that the power transferred through these links can be matched precisely to the frequency, phase, and voltage of the system to which it is connected (Masters 2004). Mai et al. (2012) estimate that 200 million MW-miles of new transmission will be needed to support a future electrical system in

Vision for a Low-Carbon-Energy Future | 239

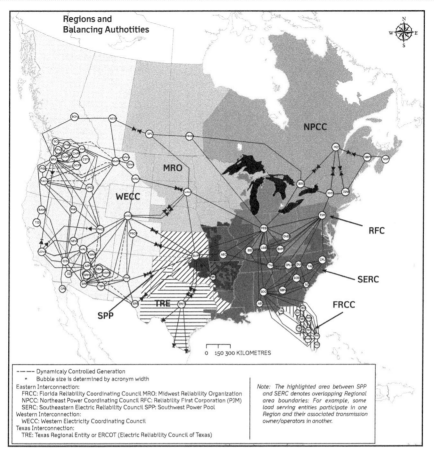

FIGURE 16.1 Regions and balancing authorities under the North American Reliability Corporation (NERC).

Notes:

Eastern Interconnection:	FRCC: Florida Reliability Coordinating Council
	MRO: Midwest Reliability Organization
	NPCC: Northeast Power Coordinating Council
	RFC: Reliability First Corporation (PJM)
	SERC: Southeastern Electric Reliability Council
	SPP: Southwest Power Pool
Western Interconnection:	WECC: Western Electricity Coordinating Council
Texas Interconnection:	TRE: Texas Regional Entity or ERCOT (Electric Reliability Council of Texas)

(*Source:* http://www.nerc.com/comm/OC/RS%20Agendas%20Highlights%20and%20Minutes%20DL/BA_Bubble_Map_20140305.jpg)

which up to 90% of electricity would be supplied from renewable sources, primarily wind and solar. Much of this investment would be concentrated in the middle and southwestern regions of the country, as indicated in Figure 16.3 (for 80% renewables in this case). Approximately 60 GW of the new capacity would be devoted to AC-DC-AC lines to

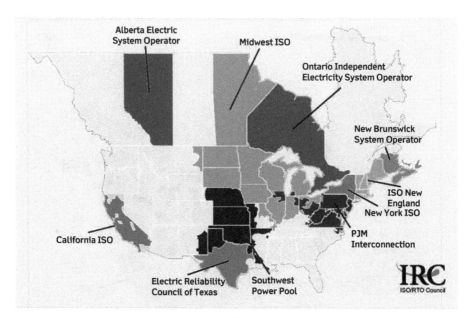

FIGURE 16.2 Independent System Operator (ISO) or regional transmission organization (RTO) regions for the United States and Canada. (Source: http://www.opuc.texas.gov/images/iso_rto_map.jpg)

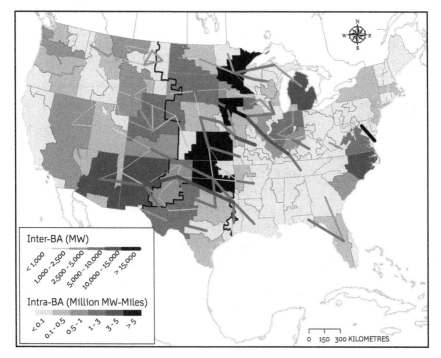

FIGURE 16.3 New transmission capacity additions required to accommodate the 80% renewable scenario considered by Mai et al. (2012). The red lines indicate additions of capacity extending across individual balancing areas with higher capacities represented by deeper shades of red. Additions within balancing areas, expressed in units of capacity miles, are indicated by the dark shading; the deeper the shading, the higher the magnitude of the relevant capacity miles.

provide for an essentially seamless connection between the present largely isolated asynchronous zones in the East, West, and Texas. Building out the system could cost as much as $300 billion, an impressive number, but not exorbitant, given that the investment could be spread over an extended period. Providing context, a 16-year history of investments in the present transmission system is presented in Figure 16.4.

Authority for siting and permitting new transmission systems in the United States rests at present with the states. Planning for and implementation of the expanded transmission network, as envisaged in Figure 16.3, will require an unprecedented commitment in terms of multistate coordination and cooperation. Not surprisingly, state regulatory agencies tend to emphasize interests of their immediate constituents. Under the circumstances, we may expect that it will be difficult, at least in some instances, for these bodies to respond affirmatively to advantages that could be realized regionally or nationally through a more comprehensive sharing of authority and interests. In the final analysis, FERC has authority to override local objections to specific multistate transmission proposals should the US Department of Energy declare them to be in the national interest. As yet, though, there is no precedent for exercise of this authority.

As discussed in Chapter 10, electricity produced from a specific wind turbine or wind farm can vary on time scales as brief as minutes or even shorter in response to

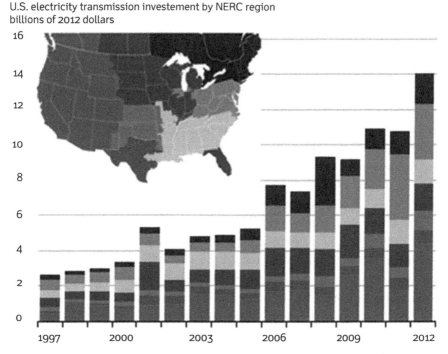

FIGURE 16.4 Annual investments in US transmission systems. Colors differentiate investments in the different North American Reliability Corporation regions as indicated. (Source: http://www.eia.gov/todayinenergy/detail.cfm?id=17811#)

fluctuations in wind speeds associated with variations in local small-scale turbulence. Huang et al. (2014) showed that this high-frequency variability could be eliminated effectively by coupling outputs from as few as 5 to 10 wind farms distributed uniformly (equally separated) over a 10-state region in the central United States. More than 95% of the residual variability would be expressed in this case on time scales longer than a day, allowing grid operators to anticipate at least a day in advance the potential supply of electricity from wind and to plan accordingly. Output of solar-generated power from a specific PV installation may be expected to vary similarly on high-frequency time scales in response to localized changes in cloud cover, more slowly on multiday time scales associated with the passage of meteorological systems, the spatial dimensions of which can extend to hundreds if not thousands of kilometers. As with the contribution from wind, combining outputs from solar sources distributed over an extended region may be expected to reduce the variability of the integrated composite source, contributing to a more reliable, more predictable power supply.

There is a further advantage that could be realized from a nationally integrated transmission system. Demand for electricity follows a typical diurnal pattern, high in the morning when people awake to begin their daily routine, with a secondary peak when they return home in the evening. The pattern in the west is delayed with respect to the east by as much as 3 hours, reflecting differences in local time. With a nationally integrated power system, inefficiencies associated with the inevitable diurnal peaks and lulls in demand could be reduced. On a national basis, demand for electricity in the United States peaks in summer, reflecting requirements for air conditioning. Wind and solar sources of electricity are complementary in the sense that the contribution from the former peaks in winter and at night, while the latter is at a maximum during the day and in summer. A nationally integrated electricity system incorporating inputs from both wind and solar could be effective in taking advantage of this synergy. There will be times though when even the combined contributions of power from wind and sun may fall short of meeting demand. There will be a need in this case for backup generation to pick up the slack.

Backup generation in the present electrical system is provided mainly by natural gas–fired systems capable of ramping up or down rapidly in response to changes in demand. Natural gas combined cycle (NGCC) systems are efficient and can respond on a time scale of hours. Gas turbines (GTs) are less efficient but have the advantage that they can ramp up almost immediately. Assuming a major future commitment to the production of electricity from wind and sun, the demand for backup generation may be expected to increase significantly, adding up to as much as a third of total generating capacity. The shortfall is likely to be greatest in summer when demand is highest and when the supply from wind is at a minimum (Mai et al. 2012). Emissions could be reduced by capturing CO_2 from the exhausts of these gas-fired plants, specifically from NGCC installations that are likely to see greater use given their higher efficiency. As discussed later, opportunities to store electricity when available in excess and to manage demand could reduce, though probably not totally eliminate, requirements for backup generation.

ENERGY USE IN THE DOMESTIC AND COMMERCIAL SECTORS

A summary of energy consumption in the US residential sector, broken down by end use, is presented in Table 16.2. Space heating accounted for 39% of the energy consumed in the sector in 2012, followed by consumption for the supply of hot water (17%), space cooling (8%), lighting (6%), refrigeration (4%), cooking (3%), TV (3%), and computers (1%). The table includes also a breakdown in the applications of energy supplied in the forms of electricity, natural gas, fuel oil, and propane. Electricity accounted for 45% of the total, followed by natural gas (41%), fuel oil (5%), and propane (5%), with the balance supplied primarily by wood products. Natural gas was responsible for 61%, 67%, and 62% of the total energy consumed in space heating, water heating, and cooking, respectively. Fuel oil and propane were used primarily for space and water heating.

Table 16.3 presents a summary of energy consumed in the commercial sector. The commercial sector classification encompasses a variety of different functions and buildings, including among others private and public offices, hotels, retail establishments, schools, and hospitals. Space heating, lighting, water heating, space cooling, and ventilation accounted for 54% of the total energy used in this sector, 50% of which was supplied by natural gas, 45% by electricity. A diversity of functions is covered under the classification of "other," including applications in hospitals and for a variety of diverse laboratory facilities. Electricity and natural gas accounted, respectively, for 45% and 41% of the total energy consumed in the commercial sector.

Initiatives at the national level have been notably successful in the United States over the past several decades in improving the energy efficiency of a wide range of commonly used electric appliances, including refrigerators, air conditioners, television sets, washing machines, computers, and lighting. Devices that meet a specified minimum standard for efficiency are given an Energy Star rating under programs administered by

TABLE 16.2

Energy Use in the US Residential Sector in 2012 (Quads)

Function	Total	Electricity	Natural Gas	Fuel Oil	Propane
Space heating	4.07	0.29	2.5	0.44	0.37
Water heating	1.79	0.45	1.2	0.05	0.07
Cooling	0.88	0.85	—	—	—
Lighting	0.64	0.64	—	—	—
Refrigerator	0.38	0.38	—	—	—
Cooking	0.34	0.11	0.21	—	0.03
TV	0.33	0.33	—	—	—
Computers	0.12	0.12	—	—	—
Total	10.42	4.69	4.26	0.51	0.51

Source: US EIA, Annual Energy Outlook 2014, http://www.eia.gov/forecasts/aeo/.

TABLE 16.3

Energy Use in the US Commercial Sector in 2012 (Quads)

Function	Total	Electricity	Natural Gas	Fuel Oil
Space heating	1.82	0.15	1.54	0.13
Lighting	0.94	0.94	—	—
Water heating	0.60	0.09	0.48	0.03
Space cooling	0.60	0.55	0.04	—
Ventilation	0.52	0.52	—	—
Refrigerator	0.38	0.38	—	—
Office equipment	0.33	0.33	—	—
Other	2.88	1.53	0.90	0.23
Total	8.29	4.52	2.96	0.42

Source: US EIA, Annual Energy Outlook 2014, http://www.eia.gov/forecasts/aeo.

the Department of Energy and the Environmental Protection Administration. When you purchase a piece of equipment, assuming that it meets minimum standards defined by these agencies, you can readily learn from the attached labeling precisely how much energy the device will consume. The program has been markedly successful in reducing electricity consumption in both the home and workplace.

In a related development, the Energy Independence and Security Act (EISA) that went into effect in December 2007 mandated that energy-inefficient light bulbs should be phased out over time, to be replaced by much more efficient alternatives. To give off light, the tungsten filament of the old standard incandescent bulb must be raised to a temperature of at least 2,500°C. Even at this elevated temperature, only 2% of the energy consumed by the light bulb is converted to photons in the visible portion of the spectrum—the light we need to see. The bulk of the energy is emitted at longer wavelengths in the infrared in the form of heat. The bulbs and lamps now on the market are notably more efficient in channeling the energy they consume into useful light. Compact fluorescent lamps (CFLs) are 3–5 times more efficient than the old incandescent lamps, and light-emitting diode bulbs (LEDs) are even better. These devices cost more, but they last longer (a function of their lower emitting temperature), and the investment pays off over a relatively brief period of time in terms of lower bills for electricity.

OPPORTUNITIES TO CONSERVE ENERGY USE
FOR SPACE HEATING AND HOT WATER

The cost for heating and cooling and for the supply of hot water accounts for a major fraction of the expense associated with the operation of both residential and commercial establishments. Investments in insulation and improved design of building shells (windows, doors, basements, walls, attics, and roofs) could provide a double dividend: reduced operational expenses and at the same time a decrease in emissions of CO_2. California has led the way in the United States in demonstrating and responding to the opportunities for savings in energy use that could be realized in the residential and commercial sectors without compromise to delivery of essential services. Statewide energy standards for buildings were instituted first in California in 1975, updated every 2 to 3 years thereafter. Consumption of natural gas per capita in the residential sector in California decreased by close to 50% between 1975 and 2005, almost twice the reduction realized over the same period in other states (Harper et al. 2011). Most of the energy savings initiatives adopted in California were implemented in the form of increasingly stringent building codes and applied primarily to new construction projects. It is more difficult to promote savings for existing structures. The decision to save or not in this case rests ultimately with property owners. How they elect to respond is likely to be guided more by considerations of personal financial advantage rather than by idealistic concerns otherwise for the health of the global environment.

A few years ago, we engaged a company to carry out an energy audit on our house in Cambridge. The audit was subsidized by the local gas/electric utility. The company that undertook this task was very professional. They sealed the doors and windows and measured the time for the air to turnover inside the house. They took infrared pictures of the interior walls and wrote a very nice report documenting their work and conclusions. In brief, they found that our house was reasonably tight but could benefit from additional insulation. I followed up, seeking to find out precisely where I should invest in this insulation, what it would cost, and how long it would take to recover the investment (the payoff). Regrettably, they were unable to provide this information, stating straightforwardly that they lacked the competence to do so. More disconcerting, they told me that they were unaware of any companies or organizations that could provide this level of end-to-end advice. If we are to encourage economically motivated decisions by owners of buildings, whether residential or commercial, to invest private capital to enhance the energy efficiency of their properties, with the attendant benefits in terms of reduced emissions of CO_2, there is clearly a need for such a service. I have a vision for how it might work.

Imagine an organization that could furnish an energy audit of a building using infrared imaging, with instrumentation mounted, for example, on a drone. These data could provide a quantitative record of precisely how much energy was leaking out of specific portions of buildings (BTUs per unit of time), and they could cover the entire neighborhood in one fell swoop. Armed with this information, the integrated energy company could supply the building owners with quantitative data on precisely how much energy was being lost from their properties and from where, what it was costing, how much capital would be required to plug the leaks, and what this would imply in terms of projected returns on investment. This could constitute, in my opinion, a win-win strategy for a broad constituency: a new business opportunity for investors; good-paying jobs for technicians who could be trained and employed to contribute productively to this important new function; savings for property owners; and, of particular relevance in the current context, an opportunity to highlight cost-effective investments that could result in significant reductions in emissions of climate-altering CO_2.

HEAT PUMPS AS AN ALTERNATIVE SOURCE OF SPACE HEAT AND HOT WATER

Meeting the overall goal for a reduction in national emissions of greenhouse gases by 83% by 2050 relative to 2005 will require contributions from the residential and commercial sectors over and above what can be realized solely by conservation. It will be important to replace a fraction of the current fossil fuel–based supply of electricity with a nonfossil alternative: as discussed earlier, wind and sun can offer plausible options. It will be necessary to cut back in addition on the use of natural gas and fuel oil as the primary energy sources deployed to heat buildings and for the supply hot water. Heat pumps fueled by low fossil fuel–based electricity could provide a constructive alternative. These systems operate by drawing energy from the external environment, either from the ambient air outside or from the ground, and deliver it to the interior of buildings in the form of heat.

Heat pumps are available in a variety of different operational modes. A particular device could function as follows. Heat would be transferred from the outside to the inside of a building through a circulating fluid. Absorption of heat by the fluid on the outside of the building could result in a change of phase of the fluid, a transition from liquid to gas. The temperature of the circulating gas could be raised subsequently by compression, with the energy for compression supplied by an electrically fueled pump. The hot gas would pass then through a series of coils, a heat exchanger, allowing heat to be transferred to the building interior either by blowing air across the coils or by pumping water across the heat exchanger. The fluid cooled in this process could condense back to the liquid state and continue its circulatory path, returning to the external exchange unit, where it would once again absorb heat from the exterior and the sequence would repeat. Traversing this cycle, the fluid would gain energy initially by transfer from the external medium (air or ground), benefit from an additional input from the compressor, delivering a fraction of the combined contributions to the interior of the building through the heat exchanger.

The efficiency of a heat pump is defined by the ratio of the electrical energy consumed in its operation to the energy content of the heat it delivers eventually to the building it services. This is referred to as the coefficient of performance (COP). The COP for a thermodynamically ideal system (a fundamental limiting theoretical construct) is determined by the ratio of the temperature at which the system absorbs heat from the outside to the temperature at which it delivers heat to the interior. The higher the external temperature and the lower the temperature at which the heat is delivered, the greater the potential value for the associated COP. Values of COPs for heat pumps currently on the market for space heating range from as low as 2 at an outdoor temperature of

0°F, increasing to about 3 for a temperature of 32°F, rising to values greater than 4 for temperatures higher than about 45°F. If the electricity used to operate the heat pumps is produced from natural gas and if there is an option to heat the building using either a natural gas–fired furnace or an electrically enabled heat pump, the breakeven point from an energy perspective is reached at a COP value of about 3 (as discussed in Chapter 2, 68% of the energy consumed in producing electricity in the United States is rejected to the environment in the form of waste heat). The breakeven price from a cost perspective depends on the difference between the prices for electricity and natural gas, which can vary significantly not only across the country but also in time.

The energy-adjusted cost for delivery of electricity to our home in Cambridge, Massachusetts, in December 2014 was 4.8 times higher than the cost for supply of natural gas. Massachusetts is unusual, however, in that the costs for electricity and natural gas are both artificially high, reflecting transportation bottlenecks in the delivery of natural gas (lack of adequate pipeline connections from supply points) and the fact that natural gas accounts for a dominant fraction of the state's production of electricity. In contrast, the residential price for electricity in Washington State in November 2014 (the state with the country's cheapest electricity) was only 2.4 times higher than the energy-adjusted price for natural gas. In terms solely of anticipated day-to-day expense, it would make sense to heat your home in Washington using an electrically powered heat pump rather than natural gas. On purely economic grounds, it would be more difficult to arrive at a similar conclusion in Massachusetts. In both states, though, switching from natural gas–fired boilers to heat pumps would result in reduced demand for energy (average winter temperatures in both states are high enough to accommodate COP values generally greater than 3), and a similar conclusion should hold for a large portion of the country. Assuming that the electricity to fuel the heat pumps was derived primarily from nonfossil sources, switching from natural gas- and oil-based sources to electrically driven heat pumps would be expected to lead to significant savings in terms of emissions of CO_2.

REDUCING EMISSIONS FROM THE TRANSPORTATION SECTOR: OPPORTUNITIES FOR ELECTRIC CARS

The transportation sector is responsible for approximately one third of US national emissions of CO_2 (Chapter 3). Emissions from this sector are associated with the use of gasoline to drive cars and light trucks; diesel oil to fuel some cars but mainly heavier trucks, ships, and trains; and jet fuel to supply requirements for domestic and military aircraft. The challenge in reducing emissions from the sector is to identify nonfossil substitutes for the fossil-based sources that currently dominate inputs of energy to this sector.

From both energy efficiency and economic perspectives, there is an advantage to using electricity rather than gasoline to drive cars and light trucks. Approximately 80% of the energy content of the gasoline consumed in a motor vehicle is wasted, converted to heat: only 20% is deployed to turn the wheels. In contrast, storing electricity in a battery and using this energy subsequently to drive the vehicle is much more efficient: losses in this instance amount to as little as 10%. Assume that 100 units of energy are used to produce 30 units of energy in the form of electricity and that 90% of this (27 units) is employed to drive a car or truck. Compare this with driving the vehicle using 100 units of energy in the form of gasoline. Only 20% of the gasoline energy (20 units) is usefully deployed in this case. Thus, driving using electricity is 25% more efficient from an energy perspective than using gasoline. From the point of view of cost, the electricity option is even more attractive.

The energy content of a gallon of gasoline is equivalent to 33.7 kWh, of which 6.7 kWh (20%) is assumed to be available to propel an internal combustion (IC) powered vehicle. The electrical energy required to deliver the driving performance equivalent to what could be supplied with a gallon of gasoline fueling an IC car or truck would amount in this case to 7.4 kWh (assuming 90% efficiency). Given a retail price of 19.8 cents per kWh for electricity in Cambridge, Massachusetts, in December 2014, it would cost $1.46 to supply the driving capacity of an electrically powered vehicle competitive with what could be achieved by consuming a gallon of gasoline in a conventional IC vehicle, a significant savings even in the context of current low prices for gasoline. The comparison would be even more impressive if applied to the market in Washington State. As indicated earlier, electricity prices there are much lower, about 9 cents per kWh (Fig. 2.1): the electrical option could provide the performance equivalence of a gallon of gasoline in this case for as little as 67 cents.

As for the relative advantages of electrically fueled transportation in terms of related emissions of CO_2, it depends on the nature of the source for the electricity. If the electricity were produced from coal, the balance would be negative: it would be preferable to use gasoline. If the electricity were produced from natural gas, there would be a significant savings in terms of emissions (production of CO_2 per unit energy delivered from natural gas is approximately half that from coal). If the electricity were produced from either wind or sun, the associated emissions would be negligible.

The market for electrically powered or electrically assisted vehicles has developed rapidly in the United States over the past few years. As of early 2015, the consumer had a choice of some 25 different models sold by some 15 different manufacturers (http://www.plugincars.com/cars). The choice includes models that operate exclusively with electricity supplied from the grid and models that use a combination of grid-supplied electricity supplemented by onboard gasoline engines deployed either to provide backup electric power or to drive the vehicles directly. The range for the electric-only mode depends on the capacity of the installed battery pack and varies from about 60 miles to as much as 265 miles. Batteries are both weighty and expensive, and prices for vehicles in this category vary accordingly. The most popular all-electric model is the Nissan Leaf priced at $29,000 with a range of 84 miles. Plug-in hybrids such as the Toyota Prius or the Honda Accord are equipped with batteries capable of drawing power from the grid and delivering between 10 and 13 of all-electric miles before switching to the standard gasoline/electric hybrid drive train. Vehicles classified as extended range have all-electric potentials that vary from as little as 20 to more than 80 miles, with options to travel much further using their gasoline backup capabilities. The 22 kWh battery on the BMW i3, for example, provides for an all-electric range of between 80 and 100 miles, which can be extended to about 190 miles using the car's onboard gasoline-powered generator. The lower cost Chevy Volt has an even greater extended-range capability, up to 300 miles, but with a much lower all-electric potential of about 38 miles.

The top of the market for the electric-only option is the Tesla Model S with a range of between 208 and 265 miles, depending on the choice of battery capacity, 60 kWh or 85 kWh. The base cost for the Tesla S runs from $64,000 to $81,000 with the lower value associated with the smaller battery option. The Model S is powered by thousands of individual lithium ion cells similar to the cells used in a variety of consumer products, including, most likely, your laptop. The battery pack is located under the floor of the car in a temperature-controlled compartment designed to ensure that the system should be secure (fireproof) even in the event of a serious collision.

As of November 2014, Tesla had installed 268 fast-charging stations around the world, with 132 in North America, capable of providing the 85 kWh Model S with an additional 150 miles of range in as little as 20 minutes. The objective of the innovative CEO of the company, Elon Musk, is that the electricity for all of these stations should be supplied eventually from the sun, from panels installed on the roofs of the structures that shelter the charging stations or located otherwise nearby. He has engaged for this purpose Solar City, the company that installed the solar panels on the roof of our house on Cape Cod (Chapter 11). Speaking to the diversity of his expertise and overlapping interests, Musk, in addition to his role as CEO at Tesla, was a cofounder and currently serves as chairman of Solar City.

Electric cars account for but a small fraction of the close to 250 million cars and light trucks on the road today in the United States. Imagine what would happen if electrically propelled vehicles should account for a much larger fraction of the total vehicle population in the future, half or even more. Consuming electricity at an average rate of about 0.3 kWh per mile (the current standard), driving 10,000 miles annually, 100 million vehicles would consume annually about 3×10^{11} kWh, adding up to close to 10% of the total current national US demand for electricity. In the process we could save 20 billion gallons of gasoline annually (assuming that gasoline-driven cars should have achieved by that time an average fuel efficiency of 50 miles per gallon) and avoid emissions of up to 176 million tons of CO_2 (assuming that the electricity for these cars and light trucks should be derived primarily from nonfossil sources). A further benefit from a large-scale transition to electrically propelled cars and light trucks is that the batteries of these vehicles could provide an important means for distributed storage of electricity.

Consumption of electricity in the United States peaks during the evening on hot days in summer with load demands ranging as high as 770 GW. Peak demand in winter is less, about 620 GW. Demand during nighttime hours when people are normally asleep drops to about 400 GW in summer, to as low as 300 GW in winter. If the fleet of electric cars envisaged here were to draw power from the grid charging their batteries over a 5-hour period at night, the additional demand for electricity would amount to about 160 GW. If the connection between the vehicles and the grid could operate in both directions, a fraction of the electric car fleet not in operation during the day could be available to supply power to the grid under conditions where demand might otherwise outstrip supply. If 30% of the fleet could meet this criterion, the related supply could amount to as much as 50 GW. The difference between day and night demands for electricity would be lowered accordingly. Fewer generating facilities would be required to meet the extra daytime demand, resulting in an increase in the efficiency of the overall electric power system. As indicated earlier, wholesale prices for electricity are typically low at night when demand is at a minimum, highest in morning and in the evening. The electric car owner could take advantage of this price differential, buying power at night when prices were low, selling during the day when prices were high. The utilities would benefit also, exploiting the flexibility afforded by the ability to store or draw power from the batteries of an expanded fleet of electric cars, thus easing the challenge faced in adjusting to the intrinsic variability of an enhanced supply of power from sun and wind.

REDUCING EMISSIONS FROM THE TRANSPORTATION SECTOR: OPTIONS FOR HEAVY VEHICLES, TRAINS, AND PLANES

Gasoline-driven cars and light trucks accounted for 60% of CO_2 emissions from the US transportation sector in 2012, with an additional 31% from diesel-powered vehicles—heavy trucks (23%), ships (5%), and trains (3%)—with a further 9% from jet fuel used in aviation. As discussed in Chapter 8, a number of the major railroads in the United States have plans to turn to natural gas in the form of LNG as a substitute for diesel to drive their trains, a development that should result in a reduction (as much as 25%) in associated emissions of CO_2. A large-scale transition from diesel to natural gas in the long-distance trucking industry is unlikely: it would require a major investment in infrastructure to accommodate demands for this alternate energy source (a supply of compressed natural gas to be available at a significant fraction of the large number of service stations that currently cater to these trucks). As discussed in Chapter 14, the source of diesel from biological sources (biodiesel) increased by 35% in 2013 relative to 2012. The increase was attributed in large measure to the introduction of a $1.00 per gallon rebate on taxes available to companies electing to blend biodiesel with conventional diesel. Despite the increase, the supply of biodiesel in the United States in 2013 amounted to less than 1.3 billion gallons, as compared with overall demand for diesel in the heavy truck sector of close to 40 billion gallons. The source of biodiesel, as currently defined and configured, is unlikely to increase significantly in the future. Despite the tax subsidy, the United States was obliged, for the first time in 2013, to turn to imports to meet requirements for the product mandated under the Renewable Fuel Standard (RFS). A better option for the future may be development of a synthetic alternative produced from plant material, taking advantage of what is known as the Fisher-Tropsch (FT) process, a procedure introduced almost 90 years ago in Germany by Franz Fisher and Hans Tropsch. Additionally, this could provide an attractive biomass-based alternative for jet fuel.

REDUCING EMISSIONS FROM THE TRANSPORTATION
SECTOR: OPPORTUNITIES FOR LOW, OR EVEN NEGATIVE,
CARBON-EMITTING FISHER-TROPSCH-DERIVED FUELS FROM BIOMASS

The FT process begins with a mixture of carbon monoxide and molecular hydrogen (CO and H_2), constituting what is referred to as synthesis gas or syngas. Syngas can be produced from a variety of feedstocks—from coal, natural gas, or biomass. It has been employed for many years in coal-rich South Africa, and earlier in similarly coal-rich Germany, to transform coal to a variety of liquid fuels, including gasoline, diesel, and jet fuel. Coal is composed of almost pure carbon. To produce the H_2 needed as input to the FT process, the CO produced from the coal is required to react with water in what is referred to as the water shift reaction. The product of this reaction, in addition to H_2, is CO_2. The quantity of CO_2 produced as an intermediate in this sequence is almost twice what is emitted to the atmosphere in conjunction with the eventual combustion of the liquid product. Assuming that the intermediate CO_2 is also emitted to the atmosphere, use of coal as a source of liquid fuel is decidedly negative in terms of its implications for net emissions of CO_2.

The FT process results in a concentrated source of CO_2. This allows for a relatively inexpensive option to capture and sequester this carbon and to bury it in a long-lived geological reservoir, effectively ensuring its permanent removal from the atmosphere. Incremental costs for capture and sequestration of CO_2 formed in the FT process are estimated to be as low as $10 to $20 a ton (Kreutz et al. 2008; Tarka et al. 2009; Schrag 2009), as compared to costs for capturing and disposing of CO_2 developed from traditional coal- and natural gas–fired power plants projected at levels of $100 a ton of CO_2 or even higher (IPCC 2005). If the CO_2 produced from FT processing of coal were captured and sequestered, emissions of CO_2 associated with combustion of the resulting liquid products would not differ significantly from emissions associated with the use of conventional liquid fossil fuel sources: 52% of the carbon contained in the coal could be captured and effectively permanently removed from the atmosphere (Kreutz et al. 2008). A more attractive option might be to employ biomass rather than coal as feedstock.

If the FT process were used to produce liquids from biomass alone, both the carbon consumed in the FT manufacturing process and the carbon emitted when the liquid product was eventually combusted would have been derived from the contemporary atmosphere. With carbon capture and sequestration, this sequence could result in a net sink for atmospheric CO_2. The downside is that the energy content of a given mass of plant material is less than the energy content of an equivalent mass of coal, resulting in a reduced potential to produce marketable fuels per mass unit of input. A more realistic option, suggested by Kreutz et al. (2008), favored also by Schrag (2009), would be to use coal but to incorporate some fraction of biomass into the initial feedstock, while capturing and sequestering the intermediate CO_2 product.

Kreutz et al. (2008) explored a variety of possible choices for feedstock and operational procedures that could be deployed to produce liquids using FT. Of particular relevance in the present context is the model identified as CBTL-RC-CCS, shorthand for conversion of coal and biomass to liquids through FT (CBTL), with recycling of unconverted syngas (RC), combined with carbon capture and sequestration (CCS). For this scenario, they postulated an input consisting of 2,441 tons of dry coal per day with 3,044 tons of dry biomass. In terms of carbon content, coal accounted for 54% of the input with 46% from biomass. Processing resulted in production of 10,000 barrels of liquid equivalent per day. The entire sequence was carbon neutral in the sense that the carbon emitted to the atmosphere when the fuel products were combusted was projected to be the same as the carbon introduced in the first place with the biomass incorporated in the feedstock. Up to 54% of the carbon supplied as input was assumed to be captured and sequestered. From an energy perspective, 44% of the energy contained in the input materials was converted to energy in the liquid products, with a further 5% deployed to generate 75 MW of carbon-free electricity available for export to the grid—an overall efficiency for end-to-end energy processing of 49%. According to the Kreutz et al. (2008) analysis, the break-even point for carbon neutrality is attained when the dry biomass content of the input material reaches a level of about 56%. With higher concentrations of biomass, the integrated sequence, from processing to fuel consumption, results in a net withdrawal of carbon from the atmosphere. The ultimate potential for biomass-fueled FT, both as a source of low-carbon transportation fuels (gasoline, diesel, and jet fuel) and as a potential sink for atmospheric CO_2, will depend on access to adequate supplies of biomass.

The advantage of the FT process applied to biomass is that, in contrast to the specific hydrocarbon inputs required for production of ethanol (corn, sugar cane, cellulose) and conventional biodiesel (soybeans, rapeseed, animal fats, recycled cooking grease, etc.), the FT process can take advantage of any available source of chemically reduced carbon. Possibilities include agricultural and municipal waste, residues from paper production, and purposely grown crops. Switchgrass offers a particularly attractive possibility. It can be grown on marginal land. It is perennial and self-seeding. Requirements for fertilizer are modest. And an acre planted to a selected hybrid form of switchgrass could produce as much as 4 tons of harvestable dry matter per year with additional carbon stored in the soil (Mitchell et al. 2013). Idle cropland in the United States amounts currently to about 40 million acres. Assuming that all of this land could be planted with switchgrass, this could provide for a sustainable annual source of as much as 160 million tons of dry biomass, sufficient, in combination with coal, to accommodate the biomass requirements for up to 155 individual plants with the 10,000 barrel per day FT carbon-neutral facilities described earlier. Liquid fuel output from these 155 plants could supply up to 18% of current (2012) US consumption of gasoline, 50% of diesel, and 150% of jet fuel. Additional supplies of biomass could be derived from fast-growing trees such as poplars and willows and could be complemented by organic-rich wastes supplied from a variety of potential industrial, agricultural, and municipal sources.

There is no doubt that biomass added to coal, processed using FT with capture and sequestration of product CO_2, could provide for a low-carbon-emitting alternative to conventional sources of gasoline, diesel, and jet fuel. At higher levels of biomass input, this could contribute to a net sink for atmospheric CO_2. As a further benefit, combining coal and biomass to produce low-carbon-emitting liquids could support a continuing market for coal, prospects for which would be limited otherwise in a fossil-constrained future world. Similar benefits in terms of production of low-carbon-emitting fuels could be realized by using natural gas rather than coal as the biomass additive (Liu et al. 2011). A key question concerns cost. A comprehensive economic analysis should consider the expense for acquisition of necessary feedstocks (coal, natural gas, and biomass), capital and operational costs for relevant processing facilities, and offsets available from marketing of the resulting liquid products. Present-day depressed domestic prices for natural gas (less than $3/MMBTU) and costs for natural gas and coal are comparable on an energy basis in the United States, suggesting that natural gas could represent an economically viable alternative to coal as an input for FT processing. Given the advantages of natural gas as a comparatively clean fuel relative to coal, however, this trend is unlikely to persist. In projecting future prices for production of liquids from FT processing of either coal or natural gas with biomass, it will be important to recognize that forecasts of this nature are intrinsically uncertain. In particular, they may fail to account for technological advances that could result in a lower cost for supply of biomass materials to a future generation of more productive FT facilities.

SUMMARY COMMENTS ON THE VISION FOR A LOW-CARBON-ENERGY FUTURE

The prospects for a low-carbon future outlined here depend on the premise that energy services will be delivered in the future to a greater extent than today through electricity. It is essential that this electricity be supplied by fossil carbon-free sources. To this end, we emphasized the potential for wind and solar. There are twin challenges that must be addressed if these sources are to live up to their potential. First, the electrical transmission system must be upgraded to allow power from favorable source locations to be transferred efficiently to regions of high demand. Second, we need to adjust to the intrinsic variability of these sources. Both objectives can be advanced by investments in an integrated national power distribution system.

Transitioning to a system in which electric companies have the ability to respond not simply to demand but also to influence this demand can play an important ancillary role in easing the challenge posed by the variability of the power source. Advances in utility-scale battery technology could be further influential. Important benefits could be derived also, as discussed, from a two-way connection between electric utilities and the batteries of a large future fleet of electrically propelled vehicles.

The guiding hand of government will be essential if we are to meet the stated US objective for emissions in 2050. Five years ago when I published my earlier book on energy (McElroy 2010), I could argue that we needed to invest in a new energy system motivated not simply by the threat of human-induced climate change but also to advance the interests of national energy security. The landscape is now totally different. The shale revolution has provided the United States with abundant sources of both oil and natural gas. Prices have plummeted. The least efficient coal-fired power plants are being mothballed and replaced by more efficient natural gas–fired plants with significantly lower operational costs. It is more difficult under these circumstances to encourage the investments needed to support the transition to the low-carbon future required to address the climate issue. There are advantages, though, to doing so, and not just in terms of minimizing damage to the climate system. The levelized costs for power generated using either wind or solar are determined almost exclusively by the initial costs of capital: the fuels are free and operational expenses are minimal. What that means is that once the initial investment is made, the cost of power is totally predictable for at least 20 years in the future. There is no comparable guarantee for power generated using fossil fuel–based alternatives.

Fossil fuels have profited from subsidies in the past and indeed continue to do so. To address the climate issue, it will be important that low-carbon-emitting alternatives should be similarly supported. Production tax credits and feed-in tariffs offer supportive options to meet this objective. It is imperative, though, that they should be implemented with a measure of predictability and continuity and that they not be subjected to the turn-on/turn-off pattern that has defined experience over much of the recent past.

ADDITIONAL PERSPECTIVES

Williams et al. (2014) discussed a number of options that could be implemented to meet President Obama's commitment for an 83% reduction in US greenhouse gas emissions by 2050 relative to 2005. Their study was conducted as part of an international initiative, the Deep Decarbonization Pathways Project (DDPP), involving teams from 15 countries responsible for 70% of total current global emissions. The goal for the DDPP is to identify strategies by which individual countries could reduce their emissions of greenhouse gases to levels consistent with limiting the future rise in global average surface temperature to 2°C or less. To this end, Williams et al. (2014) considered four options for the United States: a renewables-rich option, a nuclear-rich option, an option envisaging a major investment in carbon capture and sequestration (CCS), and a fourth incorporating a mix of the other three. The renewables-rich scenario comes closest to the vision outlined here. The stated challenge for 2050 refers to the totality of greenhouse gas emissions and not simply to CO_2 from the energy sector, as emphasized in this study. Williams et al. (2014) interpreted the composite 2050 commitment to imply that the contribution from fossil fuel–related emissions should be limited by this date to a level not to exceed 750 Mt CO_2 per year. In their study, they modeled explicitly paths they thought could be followed to meet this objective. Between 2015 and 2050, they concluded that the stock in electric lighting could turn over as many as four times. Space heating systems might be replaced twice. Industrial boilers or electric power plants, in contrast, are likely to remain in place for most of this 35-year interval, turning over at most once, except under exceptional circumstances.

Consistent with the vision elaborated here, Williams et al. (2014), in their renewables-rich scenario, assumed that electricity should play a much greater role in the future US energy system that it does today. They envisaged an increase in generating capacity from the present (2014) level of 1,070 GW to more than 3,600 GW by 2050 with the bulk of this increase taking place post 2030 as existing coal-fired plants are retired. The 2050 renewables-rich electrical system they contemplated would be dominated by contributions from wind and sun, accounting for 62.4% and 15.5% of the total, respectively. The system would be oversized to ensure that the power available would be sufficient to meet demand most of the time. When available in excess, power could be deployed to produce hydrogen, using electrolysis (the process in which water molecules are split with electricity, yielding a 2:1 mix of H_2 and O_2). The H_2 formed in this fashion could be fed directly into the natural gas distribution system, they suggested, or it could be used to produce synthetic natural gas, methane (CH_4), by reacting with CO_2—four molecules of H_2 combining with one molecule of CO_2 to produce one molecule of CH_4 and two molecules of H_2O.[2] The assumption was that this CO_2 should be derived from a biomass source and that the resulting synthetic natural gas product would be fossil carbon-free—decarbonized, to use the language favored by Williams et al. (2014). The overall sequence is referred to as power to gas (PtG). This decarbonized natural gas would be deployed, where possible, as a substitute for conventional fossil-based natural gas. In particular, it could be used to complement diesel as the fuel of choice for propulsion of heavy trucks.

Williams et al. (2014) concluded that the goal of reducing overall US greenhouse gas emissions by 83% by 2050 relative to 2005 could be met at relatively modest incremental cost. Their median estimate was for an increase in net energy system costs by 2050 of 0.8% of GDP with a 50% probability that the incremental expense should lie between −0.2% and +1.8%. They went on to note on page xii of their paper that "technology improvements and market transformation over the next decade could significantly reduce expected costs in subsequent years." Success in meeting the objectives addressed by Williams et al. (2014) and the more aspirational vision enunciated here will depend to a large extent on the posture adopted by government over the immediate future. If the challenge of human-induced climate change is taken seriously, we can chart the course to a sustainable energy future. Should we choose to ignore it, we should be prepared to live with the consequences.

CAN THE CLIMATE SYSTEM BE MANAGED TO MINIMIZE FUTURE DAMAGE?

Two possibilities have been suggested to address the issue posed by this question. The first involves altering the albedo (reflectivity) of the planet, cutting back on absorption of sunlight to offset the decreased capacity of the Earth to cool in the face of enhanced insulation resulting from the increased concentration of heat-absorbing greenhouse gases. The second proposes to remove CO_2 from the atmosphere, to bury it in a suitable long-lived depository, effectively cancelling the impact of earlier additions.

The suggestion that it might be possible to cool the Earth through active intervention is not new. The Russian climatologist Mikael Budyko proposed 40 years ago that this could be accomplished by adding sulfur to the atmosphere (Budyko 1974), complementing the impact that occurs naturally in conjunction with large volcanic eruptions. The eruption of Mount Pinatubo in the Philippines in 1991 introduced an estimated 14–26 million tons of SO_2 into the stratosphere with concentrations peaking near 25 km (Read et al. 1993). Over a period of a few months, this gaseous SO_2 was converted to sulfuric acid, forming small-sized reflective aerosols (suspended particles). This triggered transient negative radiative forcing estimated at as much as -3 W m^{-2}, comparable to the positive forcing associated with the enhanced concentration of greenhouse gases (IPCC 2007). Radiative forcing, as discussed in Chapter 4, provides a measure of the rate at which the planet is absorbing net energy from the sun over and above what is being returned to space. Air in the stratosphere turns over on a time scale of a few years in response to exchange with the underlying troposphere. The impact of a volcano on the energy budget of the atmosphere is projected thus to be relatively short-lived. The observational data indicate that the temperature of the lower atmosphere decreased by about 0.4°C in the first year following the eruption of Mount Pinatubo. Five years later, it had essentially recovered to the level that prevailed prior to the eruption (Soden et al. 2002).

Given the concerns over potentially disruptive warming triggered by the increasing concentration of greenhouse gases, it is not surprising that attention has turned more recently to the possibility that this warming could be offset through purposeful additions of sulfur to the stratosphere. The option has been identified variously as solar radiation management (Royal Society 2009), albedo modification (NAS 2015), or simply climate engineering (Keith 2013). The NAS study concluded that "albedo modification at scales sufficient to alter climate should not be deployed at this time." Keith (2013), on the other hand, expressed the view (p. xii) "that it makes sense to move with deliberate haste towards deployment of geoengineering." He contemplated an initial program in which 25,000 tons of sulfuric acid would be added to the stratosphere annually, ramping up over a decade or so to approximately 10 times this amount. He argued that in its mature phase this program could be implemented by a small number of high-flying Gulfstream business jets at an annual cost of about $750 million.

I agree with the conclusion reached in the NAS study: a commitment to large-scale albedo modifications without prior deliberate assessment would be unwise. The concern is that the remedy could turn out to be worse than the disease.

The increase in the concentration of greenhouse gases is responsible for a net gain of energy by the Earth: the planet is currently absorbing more energy from the sun that it is emitting back to space. Adding reflective material to the atmosphere would trigger assuredly an immediate decrease in the rate at which the planet is absorbing energy from the sun. The imbalance between energy in and energy out would be reduced accordingly. The problem is that this would be accomplished not by cutting back on the source of the original imbalance but rather by introducing an entirely different disturbance. To the extent that the planet is absorbing less visible light from the sun, we might expect changes to develop not just in global average temperature but also in the hydrological cycle and in the metabolism of the global biosphere. The good news is that a disturbance introduced by adding reflective material to the stratosphere would be short-lived. There would be an opportunity to access the impact in real time and to decide whether to continue with the intervention or suspend it.

There could be an ancillary benefit to global cooling from an albedo-altering intervention. The attendant decrease in atmospheric temperature would be associated most likely with a decrease in the concentration of atmospheric water vapor. As discussed in Chapter 4, the increase in the concentration of water vapor resulting from the increase in the concentration of greenhouse gases has been implicated in the extremes of weather observed in recent years—more floods, more droughts, and more energetic storms. Actions intended to mitigate the damage from weather disturbances of this nature by purposeful manipulation of the planetary albedo could be contemplated, I would suggest, but only after careful deliberation and solely if justified and supported by an extensive, focused research program.

Rather than seeking to offset the effect of one major human-induced global disturbance by introducing another, the better choice, I contend, would be to proceed with the second option noted at the beginning of this section: to remove the offending gases directly from the atmosphere, specifically the most important culprit, CO_2, and to bury it in a secure geological depository.

OPTIONS FOR CAPTURING CO_2 FROM THE ATMOSPHERE

A number of approaches have been proposed to capture CO_2 from the atmosphere. It could be removed through reaction with a solvent such as sodium hydroxide (NaOH), converted in this case to sodium carbonate (Na_2CO_3). The chemistry of the resulting solution could be modified subsequently, by addition of quick lime (CaO) for example, leading to production of a stream of concentrated CO_2 available after further treatment for transfer to a designated depository, a conveniently located deep saline aquiver, for example (IPCC 2005).

The sodium hydroxide option was proposed by Keith et al. (2005), Baciocchi et al. (2006), and Zeman (2007). Lackner (2009) opted for an alternative approach involving a commercially available plastic with the property that when dry it would be unusually effective in absorbing CO_2. The CO_2 captured in this process could be released subsequently by immersing the plastic in a bath of liquid water: CO_2 molecules absorbed by the plastic would be displaced by molecules of H_2O derived from the liquid.

There is an inevitable, thermodynamically defined minimum energy that must be expended to concentrate CO_2 from the dilute form in which it is present in the atmosphere to the much higher pressures required for processing and eventual transfer to a designated depository. Capturing the compound from the atmosphere where it is present at a partial pressure of 4×10^{-4} atm, and converting it to a concentrated stream at a pressure of 1 atm^3 requires a minimum expenditure of energy of 0.42 MMBTU per ton of CO_2. Concentrating it further to a pressure of 100 atm, as might be appropriate for transfer to a geological depository, would require an additional outlay of 0.25 MMBTU for a total of 0.67 MMBTU per ton of CO_2. The energy requirements for practical systems are likely to be significantly higher than these theoretically defined minima.

The transformations involved in capturing and concentrating CO_2 using the plastic option recommended by Lackner are intrinsically simpler than the more complex rearrangements implicit in the NaOH procedure. The energy cost should be lower accordingly, by at least a factor of 4, according to House et al. (2011). Lackner suggests that his procedure could be implemented to remove and sequester CO_2 from the atmosphere for as little as $100 per ton of CO_2 (Lackner et al. 2012). In contrast, drawing on experience with existing large-scale gas removal systems, House et al. (2011) concluded that the cost for realistic capture/sequestration systems could range as high as $1,000 per ton of CO_2. Broecker (2013), acknowledging the wide disparity between these estimates, recommends that a "major effort should be initiated to narrow the wide range of estimates currently in play" (p. 3).

To put the cost numbers in context, a cost of $100 for capturing and sequestering a ton of CO_2 from the atmosphere would be equivalent to adding 88 cents to the price of a gallon of gasoline or alternatively increasing costs for coal and natural gas by factors of 5 and 1.3, respectively, from $50 to $256 per ton in the case of coal, $4 to $5.3 per MMBTU for natural gas. Should costs range much higher than the value suggested by Lackner et al. (2012), closer to the higher limit quoted by House et al. (2011) or even the intermediate value ($600 per ton of CO_2) recommended by APS (2011), prospects for capture of CO_2 from the atmosphere by direct chemical intervention using commercially available energy would appear to be limited. Tapping energy from the sun captured by green plants through photosynthesis could provide a more promising alternative.

There are a number of approaches that could be employed to engage the biosphere in an energy-assisted application to remove carbon from the atmosphere. Biomass could be combusted to produce electricity with capture and sequestration of the resulting CO_2

(Keith et al. 2005; House et al. 2011). CO_2 could be captured and sequestered in conjunction with the production of ethanol or it could be employed to produce H_2 (Keith et al. 2005). Costs associated with any and all of these options are uncertain, with estimates for the electricity option ranging from $150 to $400 per ton of CO_2 sequestered (House et al. 2011). To these opportunities we would add the possibility of applying the FT process, as discussed earlier, to a mix of biomass and coal (or natural gas) to produce high-value liquid fuels. Combining this process with capture and sequestration of the incidentally produced CO_2, assuming a fractional abundance of more than 56% for the dry mass content of the biomass component of the input material relative to coal, could contribute to a net withdrawal of CO_2 from the atmosphere (Kreutz et al. 2008). Accounting for both the energy content of the product liquid fuels and the electricity generated as a by-product, it was concluded that up to 50% of the energy contained in the input materials (the biomass and coal in this particular instance) could be converted to energy-useful products (liquid fuels and electricity), in contrast to the much lower efficiency, closer to 20%, that would be realized if the biomass was simply combusted to produce electricity. The efficiency would be even less in this case if combined with capture and sequestration of the associated CO_2 emissions.

I offered a vision in this chapter for a future in which emissions of CO_2 associated with combustion of fossil fuels could be markedly reduced, effectively eliminated, on a time scale of decades. If this objective is to be addressed successfully, government must play an important role in guiding the necessary transition. The challenge is daunting given the current low price for fossil fuels in the United States—coal, oil, and natural gas. A tax on carbon, or alternatively a carbon-trading regime, could balance the scales, allowing renewable energy to compete with fossil alternatives. Incentives in the form of tax credits or feed-in tariffs for investment in renewable options could be similarly effective. As discussed by Williams et al. (2014), the cost to effect the transition to a low-fossil-carbon-emitting future need not be prohibitive. Indeed, there could be savings that could be realized should creativity be engaged and allowed to flourish. Public support, conviction that the threat of climate change is real and that it must be confronted, will be critical if we are to chart a successful course to the climate-friendly, low-carbon-emitting future proposed in this chapter.

KEY POINTS

1. Generation of electricity was responsible for 38% of US emissions of CO_2 in 2013 with an additional 34% derived from transportation, the balance from a combination of industrial (18%), residential (6%), and commercial (4%) sources.
2. The goal enunciated by President Obama, to reduce US emissions of greenhouse gases by 83% by 2050 relative to 2005, will require energy services to be

provided in the future to a greater extent than today in the form of electricity produced primarily from wind and sun with potential for an additional contribution from heat emanating from the Earth's interior.
3. Nuclear energy could make an important contribution to the objective of a low CO_2-emitting future. Current public antipathy and cost considerations suggest that the potential for this source of power is limited, however, for the United States at least over the near term. Prospects are more encouraging for China.
4. Electricity supplied by nonfossil sources could provide an environmentally and economically attractive substitute for gasoline and diesel oil as the energy source of choice for cars and light trucks.
5. Batteries incorporated in an expanded fleet of electrically powered vehicles could provide an important opportunity for distributed storage of electricity. Batteries of these vehicles could be charged at night when demand for electricity is low. Assuming a two-way connection between batteries and utilities, a fraction of the batteries could be available to transfer electricity to the grid under conditions where supply from conventional sources may be inadequate to keep up with demand. Access to this distributed storage could mitigate to some extent problems resulting from the variability of power supplied from sun and wind.
6. The challenge in accommodating an expanded source of variable power from wind and sun could be eased by investments in the national electrical distribution system. Improving the links between the three effectively isolated autonomous grids that define the current system could allow for more efficient transfer of power from wind-rich regions in the Midwest and sun-rich regions in the Southwest to high-demand centers in the West and East.
7. Electricity-fueled heat pumps could offer an alternative, or at least a supplement, for energy supplied currently in the form of natural gas and oil to heat water and to supply heat to commercial and residential buildings. Assuming that this electricity is produced from nonfossil sources, it could provide an additional opportunity to reduce prospective emissions of CO_2.
8. Demand for energy to heat buildings could be reduced by purposeful investments in insulation for existing structures combined with building codes to motivate more energy-efficient designs for the future.
9. The Fisher-Tropsch process could provide an opportunity for efficient conversion of biomass, in combination with either coal or natural gas, to liquid fuels, including synthetic forms of gasoline, diesel, and jet fuel. Assuming a minimum threshold for incorporation of biomass in the relevant feedstock, capture and sequestration of CO_2 evolved as a byproduct of this process could result in a net sink for atmospheric CO_2.

10. Costs to realize the goal for a low-fossil-carbon-emitting future outlined here need not be prohibitive and could be accompanied by a variety of significant ancillary benefits in terms of employment, infrastructure renewal, energy security, and international leadership. Successfully addressing the goal will require a commitment from government, supported by an educational program targeted to develop a broad-based consensus as to its importance.

Notes

1. The carbon-negative concept assumes that CO_2 emitted in conjunction with the use of these biomass-derived products should be captured and deposited in a long-lived reservoir, effectively permanently separated from the atmosphere.

2. Melaina et al. (2013) concluded that hydrogen, H_2, could be added safely to the gas distribution system with concentrations ranging potentially as high as 15% by volume. They cautioned, though, that levels judged safe and acceptable should be assessed on a case-by-case basis.

3. Pressures are quoted here in units of atmospheres, abbreviated as atm. A pressure of 1 atm is equivalent to the pressure of the atmosphere at sea level.

References

Baciocchi, R., G. Storti, et al. 2006. Process design and energy requirements for the capture of carbon dioxide from air. *Chemical Engineering and Processing* 45, no. 12: 1047–1058.

Broecker, W. 2013. Does air capture constitute a viable backstop against a bad CO_2 trip? *Elementa: Science of the Anthropocene* 1: 000009.

Budyko, M. I. 1974. *Climate and life.* New York: Academic Press.

House, K. Z., A. C. Baclig, et al. 2011. Economic and energetic analysis of capturing CO_2 from ambient air. *Proceedings of the National Academy of Sciences of the United States of America* 108, no. 51: 20428–20433.

Huang, J., X. Lu, et al. 2014. Meteorologically defined limits to reduction in the variability of outputs from a coupled wind farm system in the Central US. *Renewable Energy* 62: 331–340.

IPCC 2005. *Special report on carbon dioxide capture and storage.* Edited by B. Metz, O. Davidson, H. d. Coninck, M. Loos, and L. Meyer. Geneva, Switzerland: Intergovernmental Panel on Climate Change.

IPCC. 2007. *Climate change 2007: Mitigation of climate change.* Contribution of Working Group III to the Fourth Assessment Report of the Intergovernmental Panel on Climate Change. Edited by B. Metz, O. Davidson, P. Bosch, R. Dave, and L. Meyer, 251–322. Cambridge: The Intergovernmental Panel on Climate Change.

Keith, D. 2013. *A case for climate engineering.* Cambridge, MA: MIT Press.

Keith, D. W., M. Ha-Duong, et al. 2005. Climate strategy with CO_2 capture from the air. *Climatic Change* 74, no. 29: 17–45.

Kreutz, T. G., E. D. Larson, et al. 2008. *Fischer-Tropsch fuels from coal and biomass.* 25th Annual International Pittsburgh Coal Conference. Pittsburgh, PA: Annual International Pittsburgh Coal Conference.

Lackner, K. S. 2009. Capture of carbon dioxide from ambient air. *European Physical Journal-Special Topics* 176: 93–106.

Lackner, K. S., S. Brennan, et al. 2012. The urgency of the development of CO_2 capture from ambient air. *Proceedings of the National Academy of Sciences of the United States of America* 109, no. 33: 13156–13162.

Liu, G. J., R. H. Williams, et al. 2011. Design/economics of low-carbon power generation from natural gas and biomass with synthetic fuels co-production. *10th International Conference on Greenhouse Gas Control Technologies* 4: 1989–1996.

Mai, T., R. Wiser, et al. 2012. Volume 1: Exploration of high-penetration renewable electricity futures. In *Renewable electricity futures study*, edited by M. M. Hand, S. Baldwin, E. DeMeo, et al. Golden, CO, National Renewable Energy Laboratory.

Masters, G. M. 2004. *Renewable and efficient electric power systems*. Hoboken, NJ: John Wiley & Sons, Inc.

McElroy, M. B. 2010. *Energy perspectives, problems, and prospects*. New York: Oxford University Press.

Melaina, M. W., O. Antonia, et al. 2013. *Blending hydrogen into natural gas pipeline networks: A review of key issues*. Golden, CO: National Renewable Energy Label 131.

Mitchell, R., K. Vogel, et al. 2013. Switchgrass (Panicum virgatum) for biofuel production. Lexington, KY: eXtension 11. http://www.extension.org/pages/26635/switchgrass-panicum-virgatum-for-biofuel-production#.VQiSe8o9qPU.

NAS 2015. *Climate intervention: Reflecting sunlight to cool earth*. Washington, DC: National Academy of Sciences.

Read, W. G., L. Froidevaux, et al. 1993. Microwave limb sounder measurement of stratospheric SO_2 from the Mt. Pinatubo volcano. *Geophysical Research Letters* 20, no. 12: 1299–1302.

Royal Society. 2009. *Geoengineering the climate science, governance and uncertainty*. London: The Royal Society.

Schrag, D. 2009. Coal as a low-carbon fuel? *Nature Geoscience* 2, no. 12: 818–820.

Soden, B. J., R. T. Wetherald, et al. 2002. Global cooling after the eruption of Mount Pinatubo: A test of climate feedback by water vapor. *Science* 296, no. 5568: 727–730.

Tarka, T. J. (2009). *Affordable, low-carbon diesel fuel from domestic coal and biomass*. Washington, DC: Department of Energy.

US EIA. 2015. *Monthly energy review*. Washington, DC: US Energy Information Administration.

Williams, J. H., B. Haley, et al. 2014. *Pathways to deep decarbonization in the United States*. San Francisco: Energy and Environmental Economics (E3).

Zeman, F. 2007. Energy and material balance of CO_2 capture from ambient air. *Environmental Science & Technology* 41, no. 21: 7558–7563.

Index

aerosols, influence on climate, 43–46
Alberta, 93, 98–101
Arctic Oscillation (AO), 46
Atlantic Multidecadal Oscillation (AMO), 46

Beijing Agreement, 219–231
    challenge for China, 227–231
    challenge for the United States, 220
Bell, Alexander Graham, 6
Bernoulli's Principle, 143
Betz, Albert, 143
Betz limit, 143
biodiesel, 215–217
biomass as a substitute for oil, 205–217
Bloomberg, Michael, 1
Brazil, 32, 47, 55, 95, 98, 104, 180–184, 186, 191, 206, 207, 214–218
Broecker, Wallace, 77
Budyko, Mikael, 256
Buffett, Warren, 117, 171
Bunsen, Robert, 108
Bush, George H. W., 32, 92
Bush, George W., 92, 222

Caldeira, Ken, 134
Canada, 25, 55, 62, 79, 82, 92, 93, 94, 95, 98, 99, 100, 101, 104, 110, 116, 118, 121, 149, 180, 181, 182, 184, 185, 186, 191, 195, 203, 214, 216, 238, 240
capturing CO_2 from the atmosphere, 257–259
Carter, Jimmy, 92, 130
Chinese energy use and emissions, 30–31
Clark, Maurice, 90
climate, 34–77
    energetics, 35–38
    natural sources of variability, 46–49
    variations in the past, 8–11
climate change, human influence
    arguments to the effect that it is unimportant, 65–78
    changes in Arctic sea ice, 52
    changes in sea level, 53
    increases in extreme weather, 54–57
    increases in temperature, 43, 48–49
    shifts in Hadley Circulation, 53, 56
    why you should take it seriously, 34–62
Clinton, Hillary, 92
coal, 79–87
    consumption and production for 10 highest consuming countries (Table 6.1), 81

coal (*Cont.*)
  EPA steps to limit hazardous emissions, 86
  exports, China, 82
  imports, China, 82
  reserves (Table 6.2), 83
  source of air pollution and acid rain, 84–86
Coburn, Tom, 211
Conca, James, 210
Corker, Senator, 225
corporate average fuel economy (CAFE) standards, 222–223
Cuomo, Andrew, 1

Darby, Abraham, 89
Deng, Xiaoping, 7
Denmark, 153
Donora, Pennsylvania: smog 1948, 86

Eastern Interconnection, 238–239
Edison, Thomas, 6, 91
Einstein, Albert, 21, 23, 124, 126, 211
El Nino-Southern Oscillation (ENSO), 46
electricity price, US states, 14
Electricity Reliability Council of Texas (ERCOT), 238–239
Emanuel, Kerry, 134
Energy Independence and Security Act, 206, 222
energy units definition
  British Thermal Units (BTU), 14
  kilowatt hour, 14
  quad, 26
  therm, 14
ethanol from cellulose, 206–211
ethanol from corn, 206–211
ethanol from sugar cane, 214–215

Federal Energy Regulatory Commission (FERC), 238, 241
Feinstein, Diane, 211
Fertile Crescent, 5
Fisher, Franz, 251
Fisher-Tropsch process, 251–253
  potential with carbon sequestration to provide sink for CO_2, 252
  production of liquid fuels from combinations of fossil and biomass inputs, 251–253
Ford, Henry, 6, 13, 91, 206
Fox, Josh, 115
Friedman, Thomas, 6, 18, 54, 57, 225

Gasland documentary, 115
gasoline prices
  different countries, 18
  US states, 19
GE 2.5 MW wind turbine, 144–145
geoengineering options to compensate for human-induced change, 256–257
geothermal energy, 193–202
  enhanced geothermal technology, 199–202
  Geysers hydrothermal system, 194–195
  global perspective, 196–198
  US perspectives and prospects, 198
Germany, 18, 25, 81, 83, 95, 110, 124, 133, 145, 146, 147, 149, 156, 157, 158, 160, 165, 166, 167, 169, 175, 215, 216, 251
Glen Canyon Dam opposition, 187–188
Gould, Stephen Jay, 3, 4
greenhouse gas definition, 37
greenhouse gas historical record, 38–42

Hansen, James, 38, 72, 134
Happer, William, 70
Hartwig, Matthew A., 207
heat pump, 246–247
Helman, Christopher, 116
Hoover Dam, 179–181
humans, early history, 3–5
hydropower, 178–192
  Chinese perspective and prospects, 188–191
  global perspective, 180–187
  US perspective and prospects, 187–188

India, 6, 7, 8, 16, 19, 25, 45, 55, 62, 77, 79, 80, 81, 83, 95, 116, 128, 139, 140, 145, 146, 147, 149, 181, 189, 214
industrial revolution, 5–6
Iran, 25, 26, 91, 92, 94, 95, 103, 109, 110, 140, 149, 185, 195
Iraq, 17, 91, 92, 94, 95, 103

Japan, 7, 8, 18, 25, 79, 81, 82, 95, 110, 116, 117, 123, 124, 127, 128, 131, 132, 133, 140, 149, 165, 166, 181, 196, 197, 225
Joule, James, 14

Kemeny, John G., 130
Kennedy, Senator Edward, 151
kerosene, early application for oil, 90–91
Kerry, John G., 151

Keystone pipeline, 99–102
Kirschvink, Joseph, 9
Koch, Bill, 151

Lassiter, Joseph, 139
Lawrence Berkeley National Laboratory, 228
London smog 1952, 86

MacKay, David, 15, 21
Madden Julian Oscillation (MJO), 46
Margulis, Lynn, 3
methane, 37–42
    greenhouse gas, 37
    historical record, 39–40
    radiative forcing, 40
    sources, 42, 112
methanol, 215
Mitchell, George, 113, 115, 116
Morse, Samuel, 6
Muller, Richard, 61, 66
Murphy, Senator, 225
Musk, Elon, 249

natural gas, 107–142
    coal bed methane (CBM), 121
    consumption by country 2012, 110
    early history, 107–108
    international prices, 117
    international trade routes, 111
    production and consumption, China, 119–121
    production by country 2012, 110
    production from shale, 112–116
    prospects for US exports of LNG, 112, 118
    reserves, 109
    substitute for coal in US power sector, 118
Newcomen, Thomas, 5
nitrous oxide
    greenhouse gas, 37
    historical record, 39–40
    radiative forcing, 40
    sources, 42, 112
Nixon, Richard, 92
Norway, 110, 153, 181
nuclear accidents, 129–133
    Chernobyl, 130–131
    Fukushima-Daiichi, 131–133
    Three Mile Island, 129–130
nuclear power, 123–142

basic physics, 124–126
boiling water reactor, 127
current status, 133
future prospects, 134
reprocessing of waste, 129
treatment of waste, 128
Nuclear Regulatory Commission (NRC), 129–130

Obama, Barack, 2, 12, 31, 92, 99, 101, 102, 219, 220, 222, 227, 228, 231, 232, 254, 259
oil, 89–106
    Chinese consumption and imports, 103
    consumption by country, 95
    global resources, 94
    history of development, 90–92
    production by country, 95
    source from tar sands, 98–102
    trends in price 1861–2010, 92
    trends in production and consumption, China, 102
    US crude production and imports, 96
    US imports and exports of petroleum products, 97
options for improving the US transmission system, 237–243
Otto, Nicolas, 205

Pacific Decadal Oscillation (PDO), 46
Pew, Joseph Newton, 107, 108
Poincaré, Henri, 124
power from the sun, 160–177
    concentrated solar power systems, 171–175
    photovoltaic systems
        current status, 165–169
        manufacturing elements, 163–164
        operational elements, 161–163
        utility scale, 169–171
power from wind, 143–159
    basic physics, 143–145
    capacity historical data, 146
    Cape Wind debate, 151
    challenges posed by variability, 152–154
    current status, 145–147
    economic incentives, 155–157
    electricity production historical data, 147
    theoretical potential, 147
power generating sources, 135–138
    levelized costs, Table 9.2, 136–137
    overnight capital costs, Table 9.1, 135

Reagan, Ronald, 92
reducing emissions in the transportation sector, 247–253
reducing energy use in the residential and commercial sectors, 243–247
Rockefeller, John D., 90, 108
Romney, Mitt, 151
Roosevelt, Theodore, 91, 181
Russian Federation, 79, 81, 83, 94, 95, 256

Santayana, George, 32
Saudi Arabia, 91, 94, 95, 98, 101, 103, 104, 109, 110, 149
Spain, 5, 55, 145–147, 172–173, 216
Systeme Internationale (SI) physical units
  energy, Joule (J), 14
  length, meter (m), 14
  mass, kilogram (kg), 14
  time, second (s), 14

Thornburgh, Richard, 129
Three Gorges Dam, 189
tidal energy, 195, 202–203
Tropsch, Hans, 251

US CO_2 emissions
  breakdown in emissions in 2013, 28
  temporal trends in emissions 1949–2012, 29
US Department of Energy, 166, 188, 212, 228, 241
US energy use, 25–33, 234–262
  breakdown of uses in 2013, 17
  energy use in the US commercial sector in 2012, 244
  energy use in the US residential sector in 2012, 243
  prospects for a low-carbon future, 234–262
  temporal trends in use 1949–2012, 29
US shale plays: Antrim, Barnett, Barnett-Woodford, Eagle Ford, Fayetteville, Haynesville, Mancos, Woodford, 113

Venezuela, 91, 93, 94, 98, 101, 104, 109, 181, 182, 183, 185
vision for a low-carbon-energy future, 234–259

Walcott, C. D., 3
Watt, James, 5, 14
watt, unit of power definition, 14
Western Interconnection, 238–239
Wigley, Tom, 134
Wilson, Richard, 131